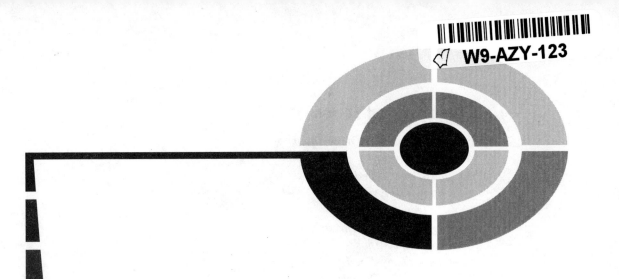

Differential Equations
Demystified

Demystified Series

Advanced Statistics Demystified
Algebra Demystified
Anatomy Demystified
Astronomy Demystified
Biology Demystified
Business Statistics Demystified
Calculus Demystified
Chemistry Demystified
College Algebra Demystified
Differential Equations Demystified
Earth Science Demystified
Electronics Demystified
Everyday Math Demystified
Geometry Demystified
Math Word Problems Demystified
Physics Demystified
Physiology Demystified
Pre-Algebra Demystified
Pre-Calculus Demystified
Project Management Demystified
Robotics Demystified
Statistics Demystified
Trigonometry Demystified

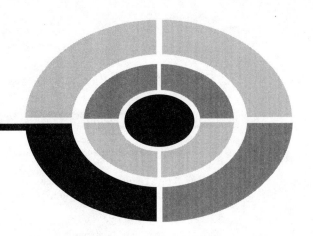

Differential Equations
Demystified

STEVEN G. KRANTZ

McGRAW-HILL
New York   Chicago   San Francisco   Lisbon   London
Madrid   Mexico City   Milan   New Delhi   San Juan
Seoul   Singapore   Sydney   Toronto

For Ed Landesman

The **McGraw·Hill** Companies

Library of Congress Cataloging-in-Publication Data

Krantz, Steven G. (Steven George), date.
    Differential equations demystified / Steven G. Krantz.
        p.   cm.
        "Demystified series" — Front matter.
        Includes bibliographical references and index.
        ISBN 0-07-144025-9 (acid-free paper)
        1. Differential equations.   I. Title.

    QA371.K63 2005
    515′.35—dc22   2004053890

3 4 5 6 7 8 9 0 DOC/DOC 0 1 0 9 8 7 6 5

ISBN 0-07-144025-9

The sponsoring editor for this book was Judy Bass and the production supervisor was Pamela A. Pelton. It was set in Times Roman by Keyword Publishing Services Ltd. The art director for the cover was Margaret Webster-Shapiro. Cover design by Handel Low.

Printed and bound by RR Donnelley.

This book was printed on recycled, acid-free paper containing a minimum of 50% recycled, de-inked fiber.

McGraw-Hill books are available at special quantity discounts to use as premiums and sales promotions, or for use in corporate training programs. For more information, please write to the Director of Special Sales, McGraw-Hill Professional, Two Penn Plaza, New York, NY 10121-2298. Or contact your local bookstore.

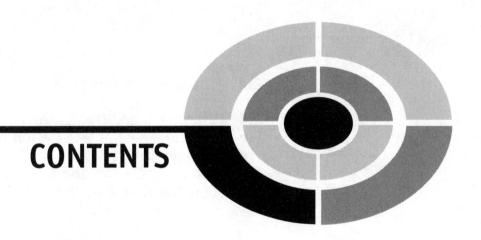

CONTENTS

Preface   ix

CHAPTER 1   **What Is a Differential Equation?**   1
1.1   Introductory Remarks   1
1.2   The Nature of Solutions   4
1.3   Separable Equations   7
1.4   First-Order Linear Equations   10
1.5   Exact Equations   13
1.6   Orthogonal Trajectories and Families
      of Curves   19
1.7   Homogeneous Equations   22
1.8   Integrating Factors   26
1.9   Reduction of Order   30
1.10 The Hanging Chain and Pursuit Curves   36
1.11 Electrical Circuits   43
      Exercises   46

CHAPTER 2   **Second-Order Equations**   48
2.1   Second-Order Linear Equations with
      Constant Coefficients   48
2.2   The Method of Undetermined
      Coefficients   54
2.3   The Method of Variation of Parameters   58

2.4   The Use of a Known Solution to
      Find Another   62
2.5   Vibrations and Oscillations   65
2.6   Newton's Law of Gravitation and
      Kepler's Laws   75
2.7   Higher-Order Linear Equations,
      Coupled Harmonic Oscillators   85
      Exercises   90

CHAPTER 3   **Power Series Solutions and
            Special Functions**   **92**
3.1   Introduction and Review of
      Power Series   92
3.2   Series Solutions of First-Order
      Differential Equations   102
3.3   Second-Order Linear Equations:
      Ordinary Points   106
      Exercises   113

CHAPTER 4   **Fourier Series: Basic Concepts**   **115**
4.1   Fourier Coefficients   115
4.2   Some Remarks About Convergence   124
4.3   Even and Odd Functions: Cosine and
      Sine Series   128
4.4   Fourier Series on Arbitrary Intervals   132
4.5   Orthogonal Functions   136
      Exercises   139

CHAPTER 5   **Partial Differential Equations and
            Boundary Value Problems**   **141**
5.1   Introduction and Historical Remarks   141
5.2   Eigenvalues, Eigenfunctions, and
      the Vibrating String   144
5.3   The Heat Equation: Fourier's
      Point of View   151
5.4   The Dirichlet Problem for a Disc   156
5.5   Sturm–Liouville Problems   162
      Exercises   166

CONTENTS

CHAPTER 6   **Laplace Transforms**   **168**
6.1   Introduction   168
6.2   Applications to Differential
      Equations   171
6.3   Derivatives and Integrals of
      Laplace Transforms   175
6.4   Convolutions   180
6.5   The Unit Step and Impulse
      Functions   189
      Exercises   196

CHAPTER 7   **Numerical Methods**   **198**
7.1   Introductory Remarks   199
7.2   The Method of Euler   200
7.3   The Error Term   203
7.4   An Improved Euler Method   207
7.5   The Runge–Kutta Method   210
      Exercises   214

CHAPTER 8   **Systems of First-Order Equations**   **216**
8.1   Introductory Remarks   216
8.2   Linear Systems   219
8.3   Homogeneous Linear Systems with
      Constant Coefficients   225
8.4   Nonlinear Systems: Volterra's
      Predator–Prey Equations   233
      Exercises   238

Final Exam   **241**

Solutions to Exercises   **271**

Bibliography   **317**

Index   **319**

PREFACE

If calculus is the heart of modern science, then differential equations are its guts. All physical laws, from the motion of a vibrating string to the orbits of the planets to Einstein's field equations, are expressed in terms of differential equations. Classically, ordinary differential equations described one-dimensional phenomena and partial differential equations described higher-dimensional phenomena. But, with the modern advent of dynamical systems theory, ordinary differential equations are now playing a role in the scientific analysis of phenomena in all dimensions.

Virtually every sophomore science student will take a course in introductory ordinary differential equations. Such a course is often fleshed out with a brief look at the Laplace transform, Fourier series, and boundary value problems for the Laplacian. Thus the student gets to see a little advanced material, and some higher-dimensional ideas, as well.

As indicated in the first paragraph, differential equations is a lovely venue for mathematical modeling and the applications of mathematical thinking. Truly meaningful and profound ideas from physics, engineering, aeronautics, statics, mechanics, and other parts of physical science are beautifully illustrated with differential equations.

We propose to write a text on ordinary differential equations that will be meaningful, accessible, and engaging for a student with a basic grounding in calculus (for example, the student who has studied *Calculus Demystified* by this author will be more than ready for *Differential Equations Demystified*). There will be many applications, many graphics, a plethora of worked examples, and hundreds of stimulating exercises. The student who completes this book will be

ready to go on to advanced analytical work in applied mathematics, engineering, and other fields of mathematical science. It will be a powerful and useful learning tool.

Steven G. Krantz

CHAPTER 1

What Is a Differential Equation?

1.1 Introductory Remarks

A *differential equation* is an equation relating some function f to one or more of its derivatives. An example is

$$\frac{d^2 f}{dx^2} + 2x \frac{df}{dx} + f^2(x) = \sin x. \tag{1}$$

Observe that this particular equation involves a function f together with its first and second derivatives. The objective in solving an equation like (1) is to *find the function f*. Thus we already perceive a fundamental new paradigm: When we solve an algebraic equation, we seek a number or perhaps a collection of numbers; but when we solve a differential equation we seek one or more *functions*.

Many of the laws of nature—in physics, in engineering, in chemistry, in biology, and in astronomy—find their most natural expression in the language of differential equations. Put in other words, differential equations are the language of nature. Applications of differential equations also abound in mathematics itself, especially in geometry and harmonic analysis and modeling. Differential equations occur in economics and systems science and other fields of mathematical science.

It is not difficult to perceive why differential equations arise so readily in the sciences. If $y = f(x)$ is a given function, then the derivative df/dx can be interpreted as the rate of change of f with respect to x. In any process of nature, the variables involved are related to their rates of change by the basic scientific principles that govern the process—that is, by the laws of nature. When this relationship is expressed in mathematical notation, the result is usually a differential equation.

Certainly Newton's Law of Universal Gravitation, Maxwell's field equations, the motions of the planets, and the refraction of light are important physical examples which can be expressed using differential equations. Much of our understanding of nature comes from our ability to solve differential equations. The purpose of this book is to introduce you to some of these techniques.

The following example will illustrate some of these ideas. According to Newton's second law of motion, the acceleration a of a body of mass m is proportional to the total force F acting on the body. The standard implementation of this relationship is

$$F = m \cdot a. \tag{2}$$

Suppose in particular that we are analyzing a falling body of mass m. Express the height of the body from the surface of the Earth as $y(t)$ feet at time t. The only force acting on the body is that due to gravity. If g is the acceleration due to gravity (about -32 ft/sec^2 near the surface of the Earth) then the force exerted on the body is $m \cdot g$. And of course the acceleration is d^2y/dt^2. Thus Newton's law (2) becomes

$$m \cdot g = m \cdot \frac{d^2 y}{dt^2} \tag{3}$$

or

$$g = \frac{d^2 y}{dt^2}.$$

We may make the problem a little more interesting by supposing that air exerts a resisting force proportional to the velocity. If the constant of proportionality is k, then the total force acting on the body is $mg - k \cdot (dy/dt)$. Then the equation (3) becomes

$$m \cdot g - k \cdot \frac{dy}{dt} = m \cdot \frac{d^2y}{dt^2}. \tag{4}$$

Equations (3) and (4) express the essential attributes of this physical system.

A few additional examples of differential equations are these:

$$(1 - x^2)\frac{d^2y}{dx^2} - 2x\frac{dy}{dx} + p(p+1)y = 0; \tag{5}$$

$$x^2\frac{d^2y}{dx^2} + x\frac{dy}{dx} + (x^2 - p^2)y = 0; \tag{6}$$

$$\frac{d^2y}{dx^2} + xy = 0; \tag{7}$$

$$(1 - x^2)y'' - xy' + p^2y = 0; \tag{8}$$

$$y'' - 2xy' + 2py = 0; \tag{9}$$

$$\frac{dy}{dx} = k \cdot y. \tag{10}$$

Equations (5)–(9) are called Legendre's equation, Bessel's equation, Airy's equation, Chebyshev's equation, and Hermite's equation respectively. Each has a vast literature and a history reaching back hundreds of years. We shall touch on each of these equations later in the book. Equation (10) is the equation of exponential decay (or of biological growth).

Math Note: A great many of the laws of nature are expressed as second-order differential equations. This fact is closely linked to Newton's second law, which expresses force as mass time acceleration (and acceleration is a *second derivative*). But some physical laws are given by higher-order equations. The Euler–Bernoulli beam equation is fourth-order.

Each of equations (5)–(9) is of second-order, meaning that the highest derivative that appears is the second. Equation (10) is of first-order, meaning that the highest derivative that appears is the first. Each equation is an *ordinary differential equation*, meaning that it involves a function of a single variable and the *ordinary derivatives* (*not* partial derivatives) of that function.

1.2 The Nature of Solutions

An ordinary differential equation of order n is an equation involving an unknown function f together with its derivatives

$$\frac{df}{dx}, \frac{d^2 f}{dx^2}, \ldots, \frac{d^n f}{dx^n}.$$

We might, in a more formal manner, express such an equation as

$$F\left(x, y, \frac{df}{dx}, \frac{d^2 f}{dx^2}, \ldots, \frac{d^n f}{dx^n}\right) = 0.$$

How do we verify that a given function f is actually the solution of such an equation?

The answer to this question is best understood in the context of concrete examples.

 EXAMPLE 1.1

Consider the differential equation

$$y'' - 5y' + 6y = 0.$$

Without saying how the solutions are actually *found*, we can at least check that $y_1(x) = e^{2x}$ and $y_2(x) = e^{3x}$ are both solutions.

To verify this assertion, we note that

$$y_1'' - 5y_1' + 6y_1 = 2 \cdot 2 \cdot e^{2x} - 5 \cdot 2 \cdot e^{2x} + 6 \cdot e^{2x}$$

$$= [4 - 10 + 6] \cdot e^{2x}$$

$$\equiv 0$$

and

$$y_2'' - 5y_2' + 6y_2 = 3 \cdot 3 \cdot e^{3x} - 5 \cdot 3 \cdot e^{3x} + 6 \cdot e^{3x}$$

$$= [9 - 15 + 6] \cdot e^{3x}$$

$$\equiv 0.$$

This process, of verifying that a *function* is a solution of the given differential equation, is most likely entirely new for you. You will want to practice and become accustomed to it. In the last example, you may check that any function of the form

$$y(x) = c_1 e^{2x} + c_2 e^{3x} \tag{1}$$

(where c_1, c_2 are arbitrary constants) is also a solution of the differential equation.

Math Note: This last observation is an instance of the principle of superposition in physics. Mathematicians refer to the algebraic operation in equation (1) as "taking a linear combination of solutions" while physicists think of the process as superimposing forces.

An important obverse consideration is this: When you are going through the procedure to solve a differential equation, how do you know when you are finished? The answer is that the solution process is complete when all derivatives have been eliminated from the equation. For then you will have y expressed in terms of x (at least implicitly). Thus you will have found the sought-after function.

For a large class of equations that we shall study in detail in the present book, we will find a number of "independent" solutions equal to the order of the differential equation. Then we will be able to form a so-called "general solution" by combining them as in (1). Of course we shall provide all the details of this process in the development below.

You Try It: Verify that each of the functions $y_1(x) = e^x$ $y_2(x) = e^{2x}$ and $y_3(x) = e^{-4x}$ is a solution of the differential equation

$$\frac{d^3y}{dx^3} + \frac{d^2y}{dx^2} - 10\frac{dy}{dx} + 8y = 0.$$

More generally, check that $y(x) = c_1e^x + c_2e^{2x} + c_3e^{-4x}$ (where c_1, c_2, c_3 are arbitrary constants) is a "general solution" of the differential equation.

Sometimes the solution of a differential equation will be expressed as an *implicitly defined function*. An example is the equation

$$\frac{dy}{dx} = \frac{y^2}{1 - xy},\qquad(2)$$

which has solution

$$xy = \ln y + c.\qquad(3)$$

Equation (3) represents a solution because all derivatives have been eliminated. Example 1.2 below contains the details of the verification that (3) is the solution of (2).

Math Note: It takes some practice to get used to the idea that an implicitly defined function is still a function. A classic and familiar example is the equation

$$x^2 + y^2 = 1.\qquad(4)$$

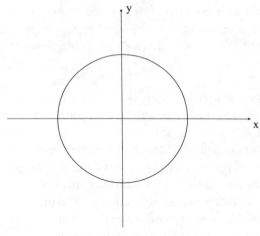

Fig. 1.1.

This relation expresses y as a function of x at most points. Refer to Fig. 1.1. In fact the equation (4) entails

$$y = +\sqrt{1 - x^2}$$

when y is positive and

$$y = -\sqrt{1 - x^2}$$

when y is negative. It is only at the exceptional points $(-1, 0)$ and $(-1, 0)$, where the tangent lines are vertical, that y *cannot* be expressed as a function of x.

Note here that the hallmark of what we call a *solution* is that it has no derivatives in it: it is a straightforward formula, relating y (the dependent variable) to x (the independent variable).

 EXAMPLE 1.2
To verify that (3) is indeed a solution of (2), let us differentiate:

$$\frac{d}{dx}[xy] = \frac{d}{dx}[\ln y + c],$$

hence

$$1 \cdot y + x \cdot \frac{dy}{dx} = \frac{dy/dx}{y}$$

or

$$\frac{dy}{dx}\left[\frac{1}{y} - x\right] = y.$$

In conclusion,

$$\frac{dy}{dx} = \frac{y^2}{1 - xy},$$

as desired.

One unifying feature of the two examples that we have now seen of verifying solutions is this: When we solve an equation of order n, we expect n "independent solutions" (we shall have to say later just what this word "independent" means) and we expect n undetermined constants. In the first example, the equation was of order 2 and the undetermined constants were c_1 and c_2. In the second example, the equation was of order 1 and the undetermined constant was c.

You Try It: Verify that the equation $x \sin y = \cos y$ gives an implicit solution to the differential equation

$$\frac{dy}{dx}[x \cot y + 1] = -1.$$

1.3 Separable Equations

In this section we shall encounter our first general class of equations with the property that

(i)   We can immediately recognize members of this class of equations.
(ii)   We have a simple and direct method for (in principle)[1] solving such equations.

This is the class of *separable equations*.

DEFINITION 1.1
An ordinary differential equation is *separable* if it is possible, by elementary algebraic manipulation, to arrange the equation so that all the dependent variables (usually the y variable) are on one side and all the independent variables

---

[1]We throw in this caveat because it can happen, and frequently does happen, that we can write down integrals that represent solutions of our differential equation, but *we are unable to evaluate those integrals*. This is annoying, but we shall later—in Chapter 7—learn numerical techniques that will address such an impasse.

(usually the x variable) are on the other side. The corresponding solution technique is called *separation of variables*.

Let us learn the method by way of some examples.

e.g.

EXAMPLE 1.3

Solve the ordinary differential equation

$$y' = 2xy.$$

SOLUTION

In the method of separation of variables—which is a method for *first-order* equations only—it is useful to write the derivative using Leibniz notation. Thus we have

$$\frac{dy}{dx} = 2xy.$$

We rearrange this equation as

$$\frac{dy}{y} = 2x\,dx.$$

[It should be noted here that we use the shorthand dy to stand for $\dfrac{dy}{dx}\,dx$.]

Now we can integrate both sides of the last displayed equation to obtain

$$\int \frac{dy}{y} = \int 2x\,dx.$$

We are fortunate in that both integrals are easily evaluated. We obtain

$$\ln y = x^2 + c.$$

[It is important here that we include the constant of integration. We combine the constant from the left-hand integral and the constant from the right-hand integral into a single constant c.] Thus

$$y = e^{x^2+c}.$$

We may abbreviate e^c by D and rewrite this last equation as

$$y = De^{x^2}. \tag{1}$$

Notice two important features of our final representation for the solution:

(i)   We have re-expressed the constant e^c as the positive constant D.
(ii) Our solution contains one free constant, as we may have anticipated since the differential equation is of order 1.

We invite you to verify that the solution in equation (1) actually satisfies the original differential equation.

EXAMPLE 1.4

Solve the differential equation

$$xy' = (1 - 2x^2) \tan y.$$

SOLUTION

We first write the equation in Leibniz notation. Thus

$$x \cdot \frac{dy}{dx} = (1 - 2x^2) \tan y.$$

Separating variables, we find that

$$\cot y \, dy = \left[\frac{1}{x} - 2x \right] dx.$$

Applying the integral to both sides gives

$$\int \cot y \, dy = \int \left[\frac{1}{x} - 2x \right] dx$$

or

$$\ln \sin y = \ln x - x^2 + C.$$

Again note that we were careful to include a constant of integration.
   We may express our solution as

$$\sin y = e^{\ln x - x^2 + C}$$

or

$$\sin y = D \cdot x \cdot e^{-x^2}.$$

The result may be written as

$$y = \sin^{-1} \left[D \cdot x \cdot e^{-x^2} \right].$$

We invite you to verify that this is indeed a solution to the given differential equation.

Math Note: It should be stressed that not all ordinary differential equations are separable. As an instance, the equation

$$x^2 y + y^2 x = \sin(xy)$$

cannot be separated so that all the x's are on one side of the equation and all the y's on the other side.

☞ **You Try It:** Use the method of separation of variables to solve the differential equation

$$x^3 y' = y.$$

1.4 First-Order Linear Equations

Another class of differential equations that is easily recognized and readily solved (at least in principle) is that of first-order linear equations.

DEFINITION 1.2
An equation is said to be *first-order linear* if it has the form

$$y' + a(x)y = b(x). \tag{1}$$

The "first-order" aspect is obvious: only first derivatives appear in the equation. The "linear" aspect depends on the fact that the left-hand side involves a differential operator that acts linearly on the space of differentiable functions. Roughly speaking, a differential equation is linear if y and its derivatives are not multiplied together, not raised to powers, and do not occur as the arguments of functions. This is an advanced idea that we shall explicate in detail later. For now, you should simply accept that an equation of the form (1) is first-order linear, and that we will soon have a recipe for solving it.

As usual, we explicate the method by proceeding directly to the examples.

 EXAMPLE 1.5
Consider the differential equation

$$y' + 2xy = x.$$

Find a complete solution.

SOLUTION

This equation is plainly not separable (try it and convince yourself that this is so). Instead we endeavor to multiply both sides of the equation by some function that will make each side readily integrable. It turns out that there is a trick that always works: You multiply both sides by $e^{\int a(x)\,dx}$.

Like many tricks, this one may seem unmotivated. But let us try it out and see how it works in practice. Now

$$\int a(x)\,dx = \int 2x\,dx = x^2.$$

[At this point we *could* include a constant of integration, but it is not necessary.] Thus $e^{\int a(x)\,dx} = e^{x^2}$. Multiplying both sides of our equation by this factor gives

$$e^{x^2} \cdot y' + e^{x^2} \cdot 2xy = e^{x^2} \cdot x$$

or

$$\left[e^{x^2} \cdot y \right]' = x \cdot e^{x^2}.$$

It is the last step that is a bit tricky. For a first-order linear equation, it is *guaranteed* that if we multiply through by $e^{\int a(x)\,dx}$ then the left-hand side of the equation will end up being the derivative of $[e^{\int a(x)\,dx} \cdot y]$. Now of course we integrate both sides of the equation:

$$\int \left[e^{x^2} \cdot y \right]' dx = \int x \cdot e^{x^2}\,dx.$$

We can perform both the integrations: on the left-hand side we simply apply the fundamental theorem of calculus; on the right-hand side we do the integration. The result is

$$e^{x^2} \cdot y = \frac{1}{2} \cdot e^{x^2} + C$$

or

$$y = \frac{1}{2} + Ce^{-x^2}.$$

Observe that, as we usually expect, the solution has one free constant (because the original differential equation was of order 1). We invite you to check that this solution actually satisfies the differential equation.

Math Note: Of course not all ordinary differential equations are first order linear. The equation

$$[y']^2 - y = \sin x$$

is indeed first order—because the highest derivative that appears is the first derivative. But it is *non*linear because the function y' is multiplied by itself. The equation

$$y'' \cdot y - y' = e^x$$

is second order and is also nonlinear—because y'' is multiplied times y.

Summary of the method of first-order linear equations

To solve a first-order linear equation

$$y' + a(x)y = b(x),$$

multiply both sides of the equation by the "integrating factor" $e^{\int a(x)\,dx}$ and then integrate.

EXAMPLE 1.6
Solve the differential equation

$$x^2 y' + xy = x^2 \cdot \sin x.$$

SOLUTION
First observe that this equation is not in the standard form (equation (1)) for first-order linear. We render it so by multiplying through by a factor of $1/x^2$. Thus the equation becomes

$$y' + \frac{1}{x}y = \sin x.$$

Now $a(x) = 1/x$, $\int a(x)\,dx = \ln|x|$, and $e^{\int a(x)\,dx} = |x|$. We multiply the differential equation through by this factor. In fact, in order to simplify the calculus, we shall restrict attention to $x > 0$. Thus we may eliminate the absolute value signs.
   Thus

$$xy' + y = x \cdot \sin x.$$

Now, as is guaranteed by the theory, we may rewrite this equation as

$$[x \cdot y]' = x \cdot \sin x.$$

Applying the integral to both sides gives

$$\int [x \cdot y]' \, dx = \int x \cdot \sin x \, dx.$$

As usual, we may use the fundamental theorem of calculus on the left, and we may apply integration by parts on the right. The result is

$$x \cdot y = -x \cdot \cos x + \sin x + C.$$

We finally find that our solution is

$$y = -\cos x + \frac{\sin x}{x} + \frac{C}{x}.$$

You should plug this answer into the differential equation and check that it works.

You Try It: Use the method of first-order linear equations to find the complete solution of the differential equation

$$y' + \frac{1}{x}y = e^x.$$

1.5 Exact Equations

A great many first-order equations may be written in the form

$$M(x, y) \, dx + N(x, y) \, dy = 0. \tag{1}$$

This particular format is quite suggestive, for it brings to mind a family of curves. Namely, if it happens that there is a function $f(x, y)$ so that

$$\frac{\partial f}{\partial x} = M \qquad \text{and} \qquad \frac{\partial f}{\partial y} = N, \tag{2}$$

then we can rewrite the differential equation as

$$\frac{\partial f}{\partial x} \, dx + \frac{\partial f}{\partial y} \, dy = 0. \tag{3}$$

Of course the only way that such an equation can hold is if

$$\frac{\partial f}{\partial x} \equiv 0 \qquad \text{and} \qquad \frac{\partial f}{\partial y} \equiv 0.$$

And this entails that the function f be identically constant. In other words,

$$f(x, y) \equiv c.$$

This last equation describes a family of curves: for each fixed value of c, the equation expresses y implicitly as a function of x, and hence gives a curve. In later parts of this book we shall learn much from thinking of the set of solutions of a differential equation as a smoothly varying family of curves in the plane.

The method of solution just outlined is called the *method of exact equations*. It depends critically on being able to tell when an equation of the form (1) can be written in the form (3). This in turn begs the question of when (2) will hold.

Fortunately, we learned in calculus a complete answer to this question. Let us review the key points. First note that, if it is the case that

$$\frac{\partial f}{\partial x} = M \qquad \text{and} \qquad \frac{\partial f}{\partial y} = N, \tag{4}$$

then we see (by differentiation) that

$$\frac{\partial^2 f}{\partial y \partial x} = \frac{\partial M}{\partial y} \qquad \text{and} \qquad \frac{\partial^2 f}{\partial x \partial y} = \frac{\partial N}{\partial x}.$$

Since mixed partials of a smooth function may be taken in any order, we find that a *necessary condition* for the condition (4) to hold is that

$$\frac{\partial M}{\partial y} = \frac{\partial N}{\partial x}. \tag{5}$$

We call (5) the *exactness condition*. This provides us with a useful test for when the method of exact equations will apply.

It turns out that condition (5) is also sufficient—at least on a domain with no holes. We refer you to any good calculus book (see, for instance, [STE]) for the details of this assertion. We will use our worked examples to illustrate the point.

 EXAMPLE 1.7
Use the method of exact equations to solve

$$\frac{x}{2} \cdot \cot y \cdot \frac{dy}{dx} = -1.$$

SOLUTION

First, we rearrange the equation as

$$2x \sin y \, dx + x^2 \cos y \, dy = 0.$$

Observe that the role of $M(x, y)$ is played by $2x \sin y$ and the role of $N(x, y)$ is played by $x^2 \cos y$. Next we see that

$$\frac{\partial M}{\partial y} = 2x \cos y = \frac{\partial N}{\partial x}.$$

Thus our necessary condition for the method of exact equations to work is satisfied. We shall soon see that it is also sufficient.

We seek a function f such that $\partial f / \partial x = M(x, y) = 2x \sin y$ and $\partial f / \partial y = N(x, y) = x^2 \cos y$. Let us begin by concentrating on the first of these conditions:

$$\frac{\partial f}{\partial x} = 2x \sin y,$$

hence

$$\int \frac{\partial f}{\partial x} \, dx = \int 2x \sin y \, dx.$$

. The left-hand side of this equation may be evaluated with the fundamental theorem of calculus. Treating x and y as independent variables (which is part of this method), we can also compute the integral on the right. The result is

$$f(x, y) = x^2 \sin y + \phi(y). \tag{6}$$

Now there is an important point that must be stressed. You should by now have expected a constant of integration to show up. But in fact our "constant of integration" is $\phi(y)$. This is because our integral was with respect to x, and therefore our constant of integration should be the most general possible expression *that does not depend on x*. That, of course, would be a function of y.

Now we differentiate both sides of (6) with respect to y to obtain

$$N(x, y) = \frac{\partial f}{\partial y} = x^2 \cos y + \phi'(y).$$

But of course we already know that $N(x, y) = x^2 \cos y$. The upshot is that

$$\phi'(y) = 0$$

or

$$\phi(y) = d,$$

an ordinary constant.

Plugging this information into equation (6) now yields that

$$f(x, y) = x^2 \sin y + d.$$

We stress that *this is not the solution of the differential equation.* Before you proceed, please review the outline of the method of exact equations that preceded this example. Our job now is to set

$$f(x, y) = c.$$

So

$$x^2 \cdot \sin y = \widetilde{c}, \tag{7}$$

where $\widetilde{c} = c - d$.

Equation (7) is in fact the solution of our differential equation, expressed implicitly. If we wish, we can solve for y in terms of x to obtain

$$y = \sin^{-1} \frac{\widetilde{c}}{x^2}.$$

And you may check that this is the solution of the given differential equation.

 EXAMPLE 1.8

Use the method of exact equations to solve the differential equation

$$y^2 \, dx - x^2 \, dy = 0.$$

SOLUTION

We first test the exactness condition:

$$\frac{\partial M}{\partial y} = 2y \neq -2x = \frac{\partial N}{\partial x}.$$

The exactness condition fails. As a result, this ordinary differential equation cannot be solved by the method of exact equations.

Notice that we are *not* saying here that the given differential equation cannot be solved. In fact it can be solved by the method of separation of variables (try it!). Rather, it cannot be solved by the method of exact equations.

Math Note: It is an interesting fact that the concept of exactness is closely linked to the geometry of the domain of the functions being studied. An important example is

$$M(x, y) = \frac{-y}{x^2 + y^2}, \qquad N(x, y) = \frac{x}{x^2 + y^2}.$$

We take the domain of M and N to be $U = \{(x,y) : 1 < x^2 + y^2 < 2\}$ in order to avoid the singularity at the origin. Of course this domain has a hole.

Then you may check that $\partial M/\partial y = \partial N/\partial x$ on U. But it can be shown that there is no function $f(x,y)$ such that $\partial f/\partial x = M$ and $\partial f/\partial y = N$. Again, the hole in the domain is the enemy.

Without advanced techniques at our disposal, it is best when using the method of exact equations to work only on domains that have no holes.

Math Note: It is a fact that, even when a differential equation fails the "exact equations test," it is always possible to multiply the equation through by an "integrating factor" so that it *will* pass the exact equations test. As an example, the differential equation

$$2xy \sin x \, dx + x^2 \sin x \, dy = 0$$

is *not* exact. But multiply through by the integrating factor $1/\sin x$ and the new equation

$$2xy \, dx + x^2 \, dy = 0$$

is exact.

Unfortunately, it can be quite difficult to discover explicitly what that integrating factor might be. We will learn more about the method of integrating factors in Section 1.8.

EXAMPLE 1.9

Use the method of exact equations to solve

$$e^y \, dx + (xe^y + 2y) \, dy = 0.$$

SOLUTION
First we check for exactness:

$$\frac{\partial M}{\partial y} = \frac{\partial}{\partial y}[e^y] = e^y = \frac{\partial}{\partial x}[xe^y + 2y] = \frac{\partial M}{\partial x}.$$

Thus the equation passes the test and the method of exact equations is at least feasible.

Now we can proceed to solve for f:

$$\frac{\partial f}{\partial x} = M = e^y,$$

hence

$$f(x, y) = x \cdot e^y + \phi(y).$$

But then

$$\frac{\partial}{\partial y} f(x, y) = \frac{\partial}{\partial y} \left[x \cdot e^y + \phi(y) \right] = x \cdot e^y + \phi'(y).$$

And this last expression must equal $N(x, y) = xe^y + 2y$. It follows that

$$\phi'(y) = 2y$$

or

$$\phi(y) = y^2 + d.$$

Altogether, then, we conclude that

$$f(x, y) = x \cdot e^y + y^2 + d.$$

We must not forget the final step. The solution of the differential equation is

$$f(x, y) = c$$

or

$$x \cdot e^y + y^2 + d = c$$

or

$$x \cdot e^y + y^2 = \tilde{c}.$$

This time we must content ourselves with the solution expressed implicitly, since it is not feasible to solve for y in terms of x.

☞ **You Try It:** Use the method of exact equations to solve the differential equation

$$3x^2 y \, dx + x^3 \, dy = 0.$$

1.6 Orthogonal Trajectories and Families of Curves

We have already noted that it is useful to think of the collection of solutions of a first-order differential equations as a family of curves. Refer, for instance, to the last example of the preceding section. We solved the differential equation

$$e^y\, dx + (xe^y + 2y)\, dy = 0$$

and found the solution set

$$x \cdot e^y + y^2 = c. \tag{1}$$

For each value of c, the equation describes a curve in the plane.

Conversely, if we are given a family of curves in the plane then we can produce a differential equation from which the curves come. Consider the example of the family

$$x^2 + y^2 = 2cx. \tag{2}$$

You can readily see that this is the family of all circles tangent to the y-axis at the origin (Fig. 1.2).

We may differentiate the equation with respect to x, thinking of y as a function of x, to obtain

$$2x + 2y \cdot \frac{dy}{dx} = 2c.$$

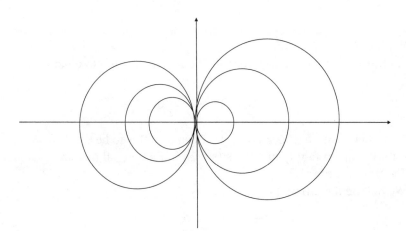

Fig. 1.2.

Now the original equation (2) tells us that

$$x + \frac{y^2}{x} = 2c,$$

and we may equate the two expressions for the quantity $2c$ (the point being to eliminate the constant c). The result is

$$2x + 2y \cdot \frac{dy}{dx} = x + \frac{y^2}{x}$$

or

$$\frac{dy}{dx} = \frac{y^2 - x^2}{2xy}. \qquad (3)$$

In summary, we see that we can pass back and forth between a differential equation and its family of solution curves.

There is considerable interest, given a family $\mathcal{F}$ of curves, to find the corresponding family $\mathcal{G}$ of curves that are orthogonal (or perpendicular) to those of $\mathcal{F}$. For instance, if $\mathcal{F}$ represents the flow curves of an electric current, then $\mathcal{G}$ will be the equipotential curves for the flow. If we bear in mind that orthogonality of curves means orthogonality of their tangents, and that orthogonality of the tangent lines means simply that their slopes are negative reciprocals, then it becomes clear what we must do.

 EXAMPLE 1.10
Find the orthogonal trajectories to the family of curves

$$x^2 + y^2 = c.$$

SOLUTION
First observe that we can differentiate the given equation to obtain

$$2x + 2y \cdot \frac{dy}{dx} = 0.$$

The constant c has disappeared, and we can take this to be the differential equation for the given family of curves (which in fact are all the circles centered at the origin—see Fig. 1.3).

We rewrite the differential equation as

$$\frac{dy}{dx} = -\frac{x}{y}.$$

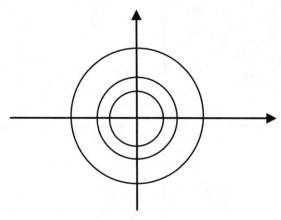

Fig. 1.3.

Now taking negative reciprocals, as indicated in the discussion right before this example, we obtain the new differential equation

$$\frac{dy}{dx} = \frac{y}{x}.$$

We may easily separate variables to obtain

$$\frac{1}{y}\, dy = \frac{1}{x}\, dx.$$

Applying the integral to both sides yields

$$\int \frac{1}{y}\, dy = \int \frac{1}{x}\, dx$$

or

$$\ln|y| = \ln|x| + C.$$

With some algebra, this simplifies to

$$|y| = D|x|$$

or

$$y = \pm Dx.$$

The solution that we have found comes as no surprise: the orthogonal trajectories to the family of circles centered at the origin is the family of lines through the origin. See Fig. 1.4.

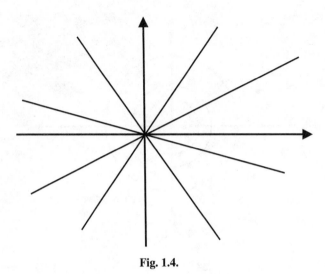

Fig. 1.4.

 Math Note: It is not the case that an "arbitrary" family of curves will have well-defined orthogonal trajectories. Consider, for example, the curves $y = |x| + c$ and think about why the orthogonal trajectories for these curves might lead to confusion.

You Try It: Find the orthogonal trajectories to the curves $y = x^2 + c$.

1.7 Homogeneous Equations

You should be cautioned that the word "homogeneous" has two meanings in this subject (as mathematics is developed simultaneously by many people all over the world, and they do not always stop to cooperate on their choices of terminology).

One usage, which we shall see and use frequently later in the book, is that an ordinary differential equation is *homogeneous* when the right-hand side is zero; that is, there is no forcing term.

The other usage will be relevant to the present section. It bears on the "balance" of weight among the different variables. It turns out that a differential equation in which the x and y variables have a balanced presence is amenable to a useful change of variables. That is what we are about to learn.

First of all, a function $g(x, y)$ of two variables is said to be *homogeneous of degree* α, for α a real number, if

$$g(tx, ty) = t^\alpha g(x, y) \qquad \text{for all } t > 0.$$

As examples, consider:

- Let $g(x, y) = x^2 + xy$. Then $g(tx, ty) = t^2 \cdot g(x, y)$, so g is homogeneous of degree 2.
- Let $g(x, y) = \sin[x/y]$. Then $g(tx, ty) = g(x, y) = t^0 \cdot g(x, y)$, so g is homogeneous of degree 0.
- Let $g(x, y) = \sqrt{x^2 + y^2}$. Then $g(tx, ty) = t \cdot g(x, y)$, so g is homogeneous of degree 1.

In case a differential equation has the form

$$M(x, y)\, dx + N(x, y)\, dy = 0$$

and M, N have the *same degree of homogeneity*, then it is possible to perform the change of variable $z = y/x$ and make the equation separable (see Section 1.3). Of course we then have a well-understood method for solving the equation.

The next examples will illustrate the method.

EXAMPLE 1.11

Use the method of homogeneous equations to solve the equation

$$(x + y)\, dx - (x - y)\, dy = 0.$$

SOLUTION

First notice that the equation is *not exact*, so we must use some other method to find a solution. Now observe that $M(x, y) = x + y$ and $N(x, y) = -(x - y)$ and each is homogeneous of degree 1. We thus rewrite the equation in the form

$$\frac{dy}{dx} = \frac{x + y}{x - y}.$$

Dividing numerator and denominator by x, we finally have

$$\frac{dy}{dx} = \frac{1 + \dfrac{y}{x}}{1 - \dfrac{y}{x}}. \tag{1}$$

The point of these manipulations is that the right-hand side is now plainly homogeneous of degree 0. We introduce the change of variable

$$z = \frac{y}{x}, \tag{2}$$

hence

$$y = zx$$

and

$$\frac{dy}{dx} = z + x \cdot \frac{dz}{dx}. \tag{3}$$

Putting (2) and (3) into (1) gives

$$z + x\frac{dz}{dx} = \frac{1+z}{1-z}.$$

Of course this may be rewritten as

$$x\frac{dz}{dx} = \frac{1+z^2}{1-z}$$

or

$$\frac{1-z}{1+z^2}\,dz = \frac{dx}{x}.$$

We apply the integral, and rewrite the left-hand side, to obtain

$$\int \frac{dz}{1+z^2} - \int \frac{z\,dz}{1+z^2} = \int \frac{dx}{x}.$$

The integrals are easily evaluated, and we find that

$$\tan^{-1} z - \frac{1}{2}\ln(1+z^2) = \ln x + C.$$

Now we return to our original notation by setting $z = y/x$. The result is

$$\tan^{-1}\frac{y}{x} - \ln\sqrt{x^2+y^2} = C.$$

Thus we have expressed y implicitly as a function of x, and thereby solved the differential equation.

 Math Note: Of course it should be clearly understood that most functions are *not* homogeneous. The functions

- $f(x, y) = x + y^2$
- $f(x, y) = x \sin y$
- $f(x, y) = e^{xy}$
- $f(x, y) = \log(x^2 y)$

have no homogeneity properties.

EXAMPLE 1.12

Solve the differential equation

$$xy' = 2x + 3y.$$

SOLUTION

It is plain that the equation is first-order linear, and we encourage the reader to solve the equation by that method for practice and comparison purposes. Instead, developing the ideas of the present section, we will use the method of homogeneous equations.

If we rewrite the equation as

$$-(2x + 3y)\, dx + x\, dy = 0,$$

then we see that each of $M = -(2x + 3y)$ and $N = x$ is homogeneous of degree 1. Thus we have as

$$\frac{dy}{dx} = \frac{2x + 3y}{x}.$$

The right-hand side is homogeneous of degree 0, as we expect.

We set $z = y/x$ and $dy/dx = z + x[dz/dx]$. The result is

$$z + x \cdot \frac{dz}{dx} = 2 + 3\frac{y}{x} = 2 + 3z.$$

The equation separates, as we anticipate, into

$$\frac{dz}{2 + 2z} = \frac{dx}{x}.$$

This is easily integrated to yield

$$\tfrac{1}{2}\ln(1 + z) = \ln x + C$$

or

$$z = Dx^2 - 1.$$

Resubstituting $z = y/x$ gives

$$\frac{y}{x} = Dx^2 - 1,$$

hence

$$y = Dx^3 - x.$$

We encourage you to check that this is indeed the solution of the given differential equation.

☞ **You Try It:**   Use the method of homogeneous equations to solve the differential equation

$$(y^2 - x^2)\,dx + xy\,dy = 0.$$

1.8 Integrating Factors

We used a special type of integrating factor in Section 1.4 on first-order linear equations. At that time, we suggested that integrating factors may be applied in some generality to the solution of first-order differential equations. The trick is in *finding* the integrating factor.

In this section we shall discuss this matter in some detail, and indicate the uses and the limitations of the method of integrating factors.

First let us illustrate the concept of integrating factor by way of a concrete example.

 EXAMPLE 1.13
The differential equation

$$y\,dx + (x^2 y - x)\,dy = 0 \tag{1}$$

is plainly *not exact*, just because $\partial M/\partial y = 1$ while $\partial N/\partial x = 2xy - 1$, and these are unequal. However, if we multiply the equation (1) through by a factor of $1/x^2$, then we obtain the equivalent equation

$$\frac{y}{x^2}\,dx + \left(y - \frac{1}{x}\right) = 0,$$

and this equation *is exact* (as you may easily verify by calculating $\partial M/\partial y$ and $\partial N/\partial x$). And of course we have a direct method (see Section 1.5) for solving such an exact equation.

We call the function $1/x^2$ in this last example an *integrating factor*. It is obviously a matter of some interest to be able to find an integrating factor for any given first-order equation. So, given a differential equation

$$M(x, y)\,dx + N(x, y)\,dy = 0,$$

we wish to find a function $\mu(x, y)$ such that

$$\mu(x, y) \cdot M(x, y)\,dx + \mu(x, y) \cdot N(x, y)\,dy = 0$$

is exact. This entails

$$\frac{\partial(\mu \cdot M)}{\partial y} = \frac{\partial(\mu \cdot N)}{\partial x}.$$

Writing this condition out, we find that

$$\mu \frac{\partial M}{\partial y} + M \frac{\partial \mu}{\partial y} = \mu \frac{\partial N}{\partial x} + N \frac{\partial \mu}{\partial x}.$$

This last equation may be rewritten as

$$\frac{1}{\mu} \left(N \frac{\partial \mu}{\partial x} - M \frac{\partial \mu}{\partial y} \right) = \frac{\partial M}{\partial y} - \frac{\partial N}{\partial x}.$$

Now we use the method of wishful thinking: we suppose not only that an integrating factor μ exists, but in fact that one exists that only depends on the variable x (and not at all on y). Then the last equation reduces to

$$\frac{1}{\mu} \frac{d\mu}{dx} = \frac{\partial M/\partial y - \partial N/\partial x}{N}.$$

Notice that the left-hand side of this new equation is a function of x only. Hence so is the right-hand side. Call the right-hand side $g(x)$. Notice that g is something that we can always compute.

Thus

$$\frac{1}{\mu} \frac{d\mu}{dx} = g(x),$$

hence

$$\frac{d(\ln \mu)}{dx} = g(x)$$

or

$$\ln \mu = \int g(x) \, dx.$$

We conclude that, in case there is an integrating factor μ that depends on x only, then

$$\mu(x) = e^{\int g(x)\, dx},$$

where

$$g(x) = \frac{\partial M}{\partial y} - \frac{\partial N}{\partial x}$$

can always be computed directly from the original differential equation.

Of course the best way to understand a new method like this is to look at some examples. This we now do.

EXAMPLE 1.14
Solve the differential equation

$$(xy - 1)\, dx + (x^2 - xy)\, dy = 0.$$

SOLUTION
You may plainly check that this equation is not exact. It is also not separable. So we shall seek an integrating factor that depends only on x. Now

$$g(x) = \frac{\partial M/\partial y - \partial N/\partial x}{N} = \frac{[x] - [2x - y]}{x^2 - xy} = -\frac{1}{x}.$$

This g depends only on x, signaling that the methodology we just developed will actually work.

We set

$$\mu(x) = e^{\int g(x)\, dx} = e^{\int -1/x\, dx} = \frac{1}{x}.$$

This is our integrating factor. We multiply the original differential equation through by $1/x$ to obtain

$$\left(y - \frac{1}{x} \right) dx + (x - y)\, dy = 0.$$

You may check that *this* equation is certainly exact. We omit the details of solving this exact equation, since that methodology was covered in Section 1.5.

Of course the roles of y and x may be reversed in our reasoning for finding an integrating factor. In case the integrating factor μ depends only on y (and not at all on x) then we set

$$h(y) = -\frac{\partial M/\partial y - \partial N/\partial x}{M}$$

and define

$$\mu(y) = e^{\int h(y)\, dy}.$$

 EXAMPLE 1.15
Solve the differential equation

$$y\, dx + (2x - ye^y)\, dy = 0.$$

SOLUTION

First observe that the equation is not exact as it stands. Second,

$$\frac{\partial M/\partial y - \partial N/\partial x}{N} = \frac{-1}{2x - ye^y}$$

does *not* depend only on x. So instead we look at

$$-\frac{\partial M/\partial y - \partial N/\partial x}{M} = -\frac{-1}{y},$$

and this expression depends only on y. So it will be our $h(y)$. We set

$$\mu(y) = e^{\int h(y)\,dy} = e^{\int 1/y\,dy} = y.$$

Multiplying the differential equation through by $\mu(y) = y$, we obtain the new equation

$$y^2\,dx + (2xy - y^2 e^y)\,dy = 0.$$

You may easily check that this new equation is exact, and then solve it by the method of Section 1.5.

You Try It: Use the method of integrating factors to transform the differential equation

$$\frac{2y}{x^2}\,dx + \frac{1}{x}\,dy = 0$$

to an exact equation. Then solve it.

Math Note: We conclude this section by noting that the differential equation

$$xy^3\,dx + yx^2\,dy = 0$$

has the properties that

- It is not exact;
- $\dfrac{\partial M/\partial y - \partial N/\partial x}{N}$ does not depend on x only;
- $-\dfrac{\partial M/\partial y - \partial N/\partial x}{M}$ does not depend on y only.

Thus the method of the present section is not a panacea. We shall not always be able to find an integrating factor. Still, the technique has its uses.

1.9 Reduction of Order

Later in the book, we shall learn that virtually *any* ordinary differential equation can be transformed to a first-order *system* of equations. This is, in effect, just a notational trick, but it emphasizes the centrality of first-order equations and systems. In the present section, we shall learn how to reduce certain higher-order equations to first-order equations—ones which we can frequently solve.

In each differential equation in this section, x will be the independent variable and y the dependent variables. So a typical second-order equation will involve x, y, y', y''. The key to the success of each of the methods that we shall introduce in this section is that one variable must be missing from the equation.

1.9.1 DEPENDENT VARIABLE MISSING

In case the variable y is missing from our differential equation, we make the substitution $y' = p$. This entails $y'' = p'$. Thus the differential equation is reduced to first-order.

 EXAMPLE 1.16
Solve the differential equation

$$xy'' - y' = 3x^2$$

using reduction of order.

SOLUTION
We set $y' = p$ and $y'' = p'$, so that the equation becomes

$$xp' - p = 3x^2.$$

Observe that this new equation is first-order linear in the new dependent variable p. We write it in standard form as

$$p' - \frac{1}{x}p = 3x.$$

We may solve this equation by using the integrating factor $\mu(x) = e^{\int -1/x \, dx} = 1/x$. Thus

$$\frac{1}{x}p' - \frac{1}{x^2}p = 3$$

so

$$\left[\frac{1}{x}p\right]' = 3$$

or

$$\int \left[\frac{1}{x}p\right]' dx = \int 3\, dx.$$

Performing the integrations, we conclude that

$$\frac{1}{x}p = 3x + C,$$

hence

$$p(x) = 3x^2 + Cx.$$

Now we recall that $p = y'$, so we make that substitution. The result is

$$y' = 3x^2 + Cx,$$

hence

$$y = x^3 + \frac{C}{2}x^2 + D = x^3 + Ex^2 + D.$$

We invite you to confirm that this is the complete and general solution to the original differential equation.

EXAMPLE 1.17

Find the solution of the differential equation

$$[y']^2 = x^2 y''.$$

SOLUTION

We note that y is missing, so we make the substitution $p = y'$, $p' = y''$. Thus the equation becomes

$$p^2 = x^2 p'.$$

This equation is amenable to separation of variables.
   The result is

$$\frac{dx}{x^2} = \frac{dp}{p^2},$$

which integrates to

$$-\frac{1}{x} = -\frac{1}{p} + E$$

or

$$p = \frac{1}{E}\left(1 - \frac{1}{1 + Ex}\right)$$

for some unknown constant E. We re-substitute $p = y'$ and integrate to obtain finally that

$$y(x) = \frac{x}{E} - \frac{1}{E^2}\ln(1 + Ex) + D$$

is the general solution of the original differential equation.

 Math Note: As usual, notice that the solution of any of our second-order differential equations gives rise to two undetermined constants. Usually these will be specified by two initial conditions.

 You Try It: Use the method of reduction of order to solve the differential equation

$$y'' - y' = x.$$

1.9.2 INDEPENDENT VARIABLE MISSING

In case the variable x is missing from our differential equation, we make the substitution $y' = p$. This time the corresponding substitution for y'' will be a bit different. To wit,

$$y'' = \frac{dp}{dx} = \frac{dp}{dy}\frac{dy}{dx} = \frac{dp}{dy} \cdot p.$$

This change of variable will reduce our differential equation to first-order. In the reduced equation, we treat p as the dependent variable (or function) and y as the independent variable.

EXAMPLE 1.18

Solve the differential equation

$$y'' + k^2 y = 0$$

[where it is understood that k is a real constant].

SOLUTION

We notice that the independent variable is missing. So we make the substitution

$$y' = p, \quad y'' = p \cdot \frac{dp}{dy}.$$

The equation then becomes

$$p \cdot \frac{dp}{dy} + k^2 y = 0.$$

In this new equation we can separate variables:

$$p \, dp = -k^2 y \, dy,$$

hence

$$\frac{p^2}{2} = -k^2 \frac{y^2}{2} + C,$$

$$p = \pm \sqrt{D - k^2 y^2} = \pm k \sqrt{E - y^2}.$$

Now we re-substitute $p = dy/dx$ to obtain

$$\frac{dy}{dx} = \pm k \sqrt{E - y^2}.$$

We can separate variables to obtain

$$\frac{dy}{\sqrt{E - y^2}} = \pm k \, dx,$$

hence

$$\sin^{-1} \frac{y}{\sqrt{E}} = \pm kx + F$$

or

$$\frac{y}{\sqrt{E}} = \sin(\pm kx + F),$$

thus

$$y = \sqrt{E} \sin(\pm kx + F).$$

Now we apply the sum formula for sine to rewrite the last expression as

$$y = \sqrt{E} \cos F \sin(\pm kx) + \sqrt{E} \sin F \cos(\pm kx).$$

A moment's thought reveals that we may consolidate the constants and finally write our general solution of the differential equation as

$$y = A \sin(kx) + B \cos(kx).$$

We shall learn in the next chapter a different, and perhaps more expeditious, method of attacking examples of the last type. It should be noted quite plainly in the last example, and also in some of the earlier examples of the section, that the method of reduction of order basically transforms the problem of solving one second-order equation to a new problem of solving *two* first-order equations. Examine each of the examples we have presented and see whether you can say what the two new equations are.

In the next example, we will solve a differential equation subject to an *initial condition*. This will be an important idea throughout the book. Solving a differential equation gives rise to a *family* of functions. Specifying the initial condition is a natural way to specialize down to a particular solution. In applications, these initial conditions will make good physical sense.

 EXAMPLE 1.19

Use the method of reduction of order to solve the differential equation

$$y'' = y' \cdot e^y$$

with initial conditions $y(0) = 0$ and $y'(0) = 1$.

SOLUTION

We make the substitution

$$y' = p, \quad y'' = p \cdot \frac{dp}{dy}.$$

So the equation becomes

$$p \cdot \frac{dp}{dy} = p \cdot e^y.$$

We of course may separate variables, so the equation becomes

$$dp = e^y \, dy.$$

This is easily integrated to give

$$p = e^y + C.$$

Now we re-substitute $p = y'$ to find that

$$y' = e^y + C$$

or

$$\frac{dy}{dx} = e^y + C.$$

Because of the initial conditions $y(0) = 0$ and $[dy/dx](0) = 1$, we may conclude right away that $C = 0$. Thus our equation is

$$\frac{dy}{e^y} = dx$$

or

$$-e^{-y} = x + D.$$

Of course we can rewrite the equation finally as

$$y = -\ln(-x + E).$$

Since $y(0) = 0$, we conclude that

$$y(x) = -\ln(-x + 1)$$

is the solution of our initial value problem.

You Try It: Use the method of reduction of order to solve the differential equation

$$y'' - y'y = 0.$$

You Try It: Use the method of reduction of order to solve the initial value problem

$$y'' + y'y = 0, \quad y(0) = 1, \quad y'(0) = 1.$$

1.10 The Hanging Chain and Pursuit Curves

1.10.1 THE HANGING CHAIN

Imagine a flexible steel chain, attached firmly at equal height at both ends, hanging under its own weight (see Fig. 1.5). What shape will it describe as it hangs?

This is a classical problem of mechanical engineering, and its analytical solution involves calculus, elementary physics, and differential equations. We describe it here.

We analyze a portion of the chain between points A and B, as shown in Fig. 1.6, where A is the lowest point of the chain and $B = (x, y)$ is a variable point. We let

- T_1 be the horizontal tension at A;
- T_2 be the component of tension *tangent* to the chain at B;
- w be the weight of the chain per unit of length.

Here T_1, T_2, w are numbers. Figure 1.7 exhibits these quantities.

Notice that if s is the length of the chain between two given points, then sw is the downward force of gravity on this portion of the chain; this is indicated in the figure. We use the symbol θ to denote the angle that the tangent to the chain at B makes with the horizontal.

By Newton's first law we may equate horizontal components of force to obtain

$$T_1 = T_2 \cos \theta. \tag{1}$$

Likewise, we equate vertical components of force to obtain

$$ws = T_2 \sin \theta. \tag{2}$$

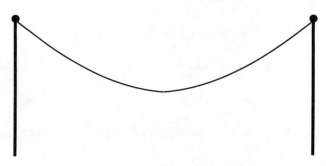

Fig. 1.5.

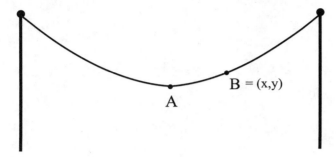

Fig. 1.6.

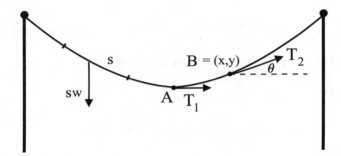

Fig. 1.7.

Dividing the right side of (2) by the right side of (1) and the left side of (2) by the left side of (1) and equating gives

$$\frac{ws}{T_1} = \tan \theta.$$

Think of the hanging chain as the graph of a function: y is a function of x. Then y' at B equals $\tan \theta$, so we may rewrite the last equation as

$$y' = \frac{ws}{T_1}.$$

We can simplify this equation by a change of notation: set $q = y'$. Then we have

$$q(x) = \frac{w}{T_1} s(x). \qquad (3)$$

If Δx is an increment of x, then $\Delta q = q(x + \Delta x) - q(x)$ is the corresponding increment of q and $\Delta s = s(x + \Delta x) - s(x)$ the increment in s. As Fig. 1.8

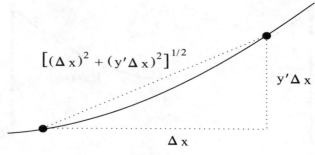

$$\left[(\Delta x)^2 + (y'\Delta x)^2\right]^{1/2}$$

$y'\Delta x$

Δx

Fig. 1.8.

indicates, Δs is well approximated by

$$\Delta s \approx \left((\Delta x)^2 + (y'\Delta x)^2\right)^{1/2} = \left(1 + (y')^2\right)^{1/2}\Delta x = (1 + q^2)^{1/2}\Delta x.$$

Thus, from (3), we have

$$\Delta q = \frac{w}{T_1}\Delta s \approx \frac{w}{T_1}(1 + q^2)^{1/2}\Delta x.$$

Dividing by Δx and letting Δx tend to zero gives the equation

$$\frac{dq}{dx} = \frac{w}{T_1}(1 + q^2)^{1/2}. \qquad (4)$$

This may be rewritten as

$$\int \frac{dq}{(1 + q^2)^{1/2}} = \frac{w}{T_1}\int dx.$$

It is trivial to perform the integration on the right side of the equation, and a little extra effort enables us to integrate the left side (use the substitution $u = \tan \psi$, or else use inverse hyperbolic trigonometric functions). Thus we obtain

$$\sinh^{-1} q = \frac{w}{T_1}x + C.$$

We know that the chain has a horizontal tangent when $x = 0$ (this corresponds to the point A—Fig. 1.7). Thus $q(0) = y'(0) = 0$. Substituting this into the last equation gives $C = 0$. Thus our solution is

$$\sinh^{-1} q(x) = \frac{w}{T_1}x$$

or

$$q(x) = \sinh\left(\frac{w}{T_1}x\right)$$

or

$$\frac{dy}{dx} = \sinh\left(\frac{w}{T_1}x\right).$$

Finally, we integrate this last equation to obtain

$$y(x) = \frac{T_1}{w}\cosh\left(\frac{w}{T_1}x\right) + D,$$

where D is a constant of integration. The constant D can be determined from the height h_0 of the point A from the x-axis:

$$h_0 = y(0) = \frac{T_1}{w}\cosh(0) + D,$$

hence

$$D = h_0 - \frac{T_1}{w}.$$

Our hanging chain is completely described by the equation

$$y(x) = \frac{T_1}{w}\cosh\left(\frac{w}{T_1}x\right) + h_0 - \frac{T_1}{w}.$$

This curve is called a *catenary*, from the Latin word for chain (*catena*). Catenaries arise in a number of other physical problems, including the *brachistochrone* and *tautochrone* which are discussed in this book. The St. Louis arch is in the shape of a catenary.

Math Note: The brachistochrone and tautochrone are discussed further in Section 6.4. These are important problems in the history of mathematics and mechanics. The brachistochrone asks for the curve of quickest descent between two given points. The tautochrone asks for a curve with the property that a bead sliding down the curve will reach bottom in the same amount of time—no matter from which height it is released. Johann Bernoulli and Isaac Newton played decisive roles in the solutions of these problems.

1.10.2 PURSUIT CURVES

A submarine speeds across the ocean bottom in a particular path, and a destroyer at a remote location decides to engage in pursuit. What path does the destroyer follow? Problems of this type are of interest in a variety of applications. We examine a few examples. The first one is purely mathematical, and devoid of "real world" trappings.

 EXAMPLE 1.20

A point P is dragged along the x–y plane by a string PT of fixed length a. If T begins at the origin and moves along the positive y-axis, and if P starts at the point $(a, 0)$, then what is the path of P?

SOLUTION

The curve described by P is called, in the classical literature, a *tractrix* (from the Latin *tractum*, meaning "drag"). Figure 1.9 exhibits the salient features of the problem.

Observe that we can calculate the slope of the pursuit curve at the point P in two ways: (i) as the derivative of y with respect to x, and (ii) as the ratio of sides of the relevant triangle. This leads to the equation

$$\frac{dy}{dx} = -\frac{\sqrt{a^2 - x^2}}{x}.$$

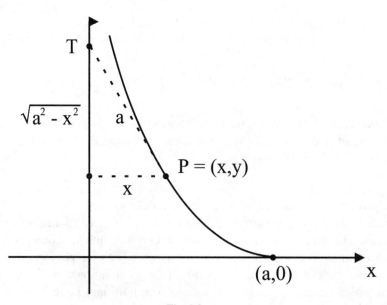

Fig. 1.9.

This is a separable, first-order differential equation. We write

$$\int dy = -\int \frac{\sqrt{a^2 - x^2}}{x}\, dx.$$

Performing the integrations (the right-hand side requires the trigonometric substitution $x = \sin \psi$), we find that

$$y = a \ln \left(\frac{a + \sqrt{a^2 - x^2}}{x} \right) - \sqrt{a^2 - x^2}$$

is the equation of the tractrix.[2]

EXAMPLE 1.21

A rabbit begins at the origin and runs up the y-axis with speed a feet per second. At the same time, a dog runs at speed b from the point $(c, 0)$ in pursuit of the rabbit. What is the path of the dog?

SOLUTION

At time t, measured from the instant both the rabbit and the dog start, the rabbit will be at the point $R = (0, at)$ and the dog at $D = (x, y)$. We wish to solve for y as a function of x. Refer to Fig. 1.10.

The premise of a pursuit analysis is that the line through D and R is tangent to the path—that is, the dog will always run straight at the rabbit. This immediately gives the differential equation

$$\frac{dy}{dx} = \frac{y - at}{x}.$$

This equation is a bit unusual for us, since x and y are both unknown functions of t. First, we rewrite the equation as

$$xy' - y = -at.$$

[Here the $'$ on y stands for differentiation in x.] We differentiate this equation with respect to x, which gives

$$xy'' = -a\frac{dt}{dx}.$$

---

[2]This curve is of considerable interest in other parts of mathematics. If it is rotated about the y-axis, then the result is a surface that gives a model for non-Euclidean geometry. The surface is called a *pseudosphere* in differential geometry. It is a surface of constant negative curvature (as opposed to a traditional sphere, which is a surface of constant positive curvature).

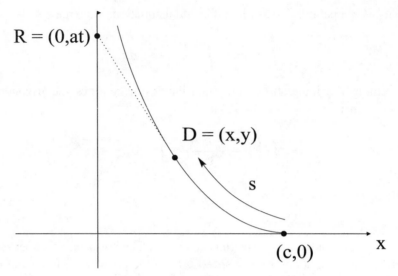

Fig. 1.10.

Since s is arc length along the path of the dog, it follows that $ds/dt = b$. Hence

$$\frac{dt}{dx} = \frac{dt}{ds} \cdot \frac{ds}{dx} = -\frac{1}{b} \cdot \sqrt{1 + (y')^2};$$

here the minus sign appears because s decreases when x increases (see Fig. 1.10). Combining the last two displayed equations gives

$$sy'' = \frac{a}{b}\sqrt{1 + (y')^2}.$$

For convenience, we set $k = a/b$, $y' = p$, and $y'' = dp/dx$ (the latter two substitutions being one of our standard reduction of order techniques). Thus we have

$$\frac{dp}{\sqrt{1 + p^2}} = k\frac{dx}{x}.$$

Now we may integrate, using the condition $p = 0$ when $x = c$. The result is

$$\ln\left(p + \sqrt{1 + p^2}\right) = \ln\left(\frac{x}{c}\right)^k.$$

When we solve for p, we find that

$$\frac{dy}{dx} = p = \frac{1}{2}\left[\left(\frac{x}{c}\right)^k - \left(\frac{c}{x}\right)^k\right].$$

In order to continue the analysis, we need to know something about the relative sizes of a and b. Suppose, for example, that $a < b$ (so $k < 1$), meaning that the dog will certainly catch the rabbit. Then we can integrate the last equation to obtain

$$y(x) = \frac{1}{2}\left[\frac{c}{k+1}\left(\frac{x}{c}\right)^{k+1} - \frac{c}{(1-k)}\left(\frac{c}{x}\right)^{k-1}\right] + D.$$

Since $y = 0$ when $x = c$, we find that $D = ck$. Of course the dog catches the rabbit when $x = 0$. Since both exponents on x are positive, we can set $x = 0$ and solve for y to obtain $y = ck$ as the point at which the dog and the rabbit meet.

We invite you to consider what happens when $a = b$ and hence $k = 1$.

Math Note: The idea of and analysis of pursuit curves is of great interest to the navy. Battle strategies are devised using these ideas.

1.11 Electrical Circuits

We have alluded elsewhere in the book to the fact that our analyses of vibrating springs and other mechanical phenomena are analogous to the situation for electrical circuits. Now we shall examine this matter in some detail.

We consider the flow of electricity in the simple electrical circuit exhibited in Fig. 1.11. The elements that we wish to note are these:

A. A source of electromotive force (emf) E—perhaps a battery or generator—which drives electric charge and produces a current I. Depending on the nature of the source, E may be a constant or a function of time.

B. A resistor of resistance R, which opposes the current by producing a drop in emf of magnitude

$$E_R = RI.$$

This equation is called *Ohm's Law*.

C. An inductor of inductance L, which opposes any change in the current by producing a drop in emf of magnitude

$$E_L = L \cdot \frac{dI}{dt}.$$

D. A capacitor (or condenser) of capacitance C, which stores the charge Q. The charge accumulated by the capacitor resists the inflow of additional charge,

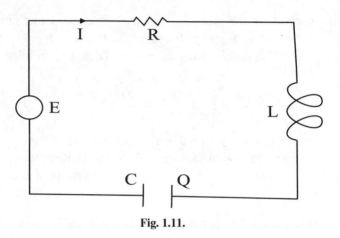

Fig. 1.11.

and the drop in emf arising in this way is

$$E_C = \frac{1}{C} \cdot Q.$$

Furthermore, since the current is the rate of flow of charge, and hence the rate at which charge builds up on the capacitor, we have

$$I = \frac{dQ}{dt}.$$

Those unfamiliar with the theory of electricity may find it helpful to draw an analogy here between the current I and the rate of flow of water in a pipe. The electromotive force E plays the role of a pump producing pressure (voltage) that causes the water to flow. The resistance R is analogous to friction in the pipe—which opposes the flow by producing a drop in the pressure. The inductance L is a sort of inertia that opposes any change in flow by producing a drop in pressure if the flow is increasing and an increase in pressure if the flow is decreasing. To understand this last point, think of a cylindrical water storage tank that the liquid enters through a hole in the bottom. The deeper the water in the tank (Q), the harder it is to pump new water in; and the large the base of the tank (C) for a given quantity of stored water, the shallower is the water in the tank and the easier to pump in new water.

The four circuit elements act together according to *Kirchhoff's Law*, which states that the algebraic sum of the electromotive forces around a closed circuit is zero. This physical principle yields

$$E - E_R - E_L - E_C = 0$$

or

$$E - RI - L\frac{dI}{dt} - \frac{1}{C}Q = 0,$$

which we rewrite in the form

$$L\frac{dI}{dt} + RI + \frac{1}{C}Q = E. \tag{1}$$

We may perform our analysis by regarding either the current I or the charge Q as the dependent variable (obviously time t will be the independent variable).

- In the first instance, we shall eliminate the variable Q from (1) by differentiating the equation with respect to t and replacing dQ/dt by I (since current is indeed the rate of change of charge). The result is

$$L\frac{d^2I}{dt^2} + R\frac{dI}{dt} + \frac{1}{C}I = \frac{dE}{dt}.$$

- In the second instance, we shall eliminate the I by replacing it by dQ/dt. The result is

$$L\frac{d^2Q}{dt^2} + R\frac{dQ}{dt} + \frac{1}{C}Q = E. \tag{2}$$

Both these ordinary differential equations are second-order linear with constant coefficients. We shall study these in detail in Section 2.1. For now, in order to use the techniques we have already learned, we assume that our system has no capacitor present. Then the equation becomes

$$L\frac{dI}{dt} + RI = E. \tag{3}$$

EXAMPLE 1.22

Solve equation (3) when an initial current I_0 is flowing and a constant emf E_0 is impressed on the circuit at time $t = 0$.

SOLUTION
For $t \geq 0$ our equation is

$$L\frac{dI}{dt} + RI = E_0.$$

We can separate variables to obtain

$$\frac{dI}{E_0 - RI} = \frac{1}{L}dt.$$

We integrate and use the initial condition $I(0) = I_0$ to obtain

$$\ln(E_0 - RI) = -\frac{R}{L}t + \ln(E_0 - RI_0),$$

hence

$$I = \frac{E_0}{R} + \left(I_0 - \frac{E_0}{R}\right)e^{-Rt/L}.$$

We have learned that the current I consists of a *steady-state* component E_0/R and a *transient component* $(I_0 - E_0/R)e^{-Rt/L}$ that approaches zero as $t \to +\infty$. Consequently, Ohm's Law $E_0 = RI$ is nearly true for t large. We also note that if $I_0 = 0$, then

$$I = \frac{E_0}{R}(1 - e^{-Rt/L});$$

if instead $E_0 = 0$, then $I = I_0 e^{-Rt/L}$.

Exercises

1. Verify that the following functions (explicit or implicit) are solutions of the corresponding differential equations:
   (a) $y = x^2 + c$ $\qquad\qquad\qquad$ $y' = 2x$
   (b) $y = cx^2$ $\qquad\qquad\qquad\quad$ $xy' = 2y$

2. Find the general solution of each of the following differential equations:
   (a) $y' = e^{3x} - x$
   (b) $y' = xe^{x^2}$

3. For each of the following differential equations, find the particular solution that satisfies the given initial condition:
   (a) $y' = xe^x$ $\qquad\qquad\qquad$ $y = 3$ when $x = 1$
   (b) $y' = 2\sin x \cos x$ $\qquad\qquad$ $y = 1$ when $x = 0$

4. Use the method of separation of variables to solve each of these ordinary differential equations:
   (a) $x^5 y' - y^{-5} = 0$
   (b) $y' = 4xy$

5. For each of the following differential equations, find the particular solution that satisfies the additional given property (called an *initial condition*):
    (a) $y'y = x + 1$ $y = 3$ when $x = 1$
    (b) $(dy/dx)x^2 = y$ $y = 2$ when $x = 1$

6. Find the general solution of each of the following first-order, linear ordinary differential equations:
    (a) $y' - xy = 0$
    (b) $y' + 2xy = 2x$

7. A tank contains 10 gallons of brine in which 2 pounds of salt are dissolved. New brine containing 1 pound of salt per gallon is pumped into the tank at the rate of 3 gallons per minute. The mixture is stirred and drained off at the rate of 4 gallons per minute. Find the amount $x = x(t)$ of salt in the tank at any time t.

8. Determine which of the following equations is exact. Solve those that *are* exact by using the method of exact equations.
    (a) $\left(x + \dfrac{2}{y} \right) dy + y\, dx = 0$

    (b) $(\sin x \tan y + 1)\, dx - \cos x \sec^2 y\, dy = 0$

9. What are the orthogonal trajectories of the family of curves $y = cx^4$?

10. Verify that each of the following equations is homogeneous, and then solve it:
    (a) $x \left(\sin \dfrac{y}{x} \right) \dfrac{dy}{dx} = y \sin \dfrac{y}{x} + x$
    (b) $xy' = y + 2xe^{-y/x}$

11. Solve each of the following differential equations by finding an integrating factor:
    (a) $12yx^2\, dx + 12x^3\, dy = 0$
    (b) $(xy - 1)\, dx + (x^2 - xy)\, dy = 0$

12. Find a solution to each of the following differential equations using the method of reduction of order:
    (a) $xy'' = y' + (y')^3$
    (b) $y'' - k^2 y = 0$

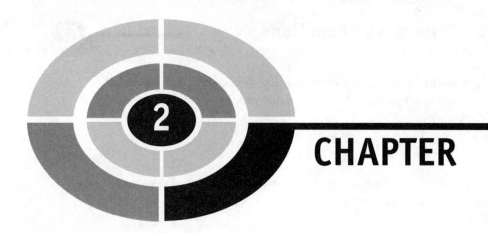

CHAPTER 2

Second-Order Equations

2.1 Second-Order Linear Equations with Constant Coefficients

Second-order linear equations are important because they arise frequently in engineering and physics. For instance, acceleration is given by the second derivative, and force is mass times acceleration.

In this section we learn about *second-order linear equations with constant coefficients*. The "linear" attribute means, just as it did in the first-order situation, that the unknown function and its derivatives are not multiplied together, are not raised to powers, and are not the arguments of other function. So, for example,

$$y'' - 3y' + 6y = 0$$

is second-order linear while

$$\sin y'' - y' + 5y = 0$$

and

$$y \cdot y'' + 4y' + 3y = 0$$

are not.

The "constant coefficient" attribute means that the coefficients in the equation are not functions—they are constants. Thus a second-order linear equation with constant coefficient will have the form

$$ay'' + by' + cy = d, \tag{1}$$

where a, b, c, d are constants.

We in fact begin with the *homogeneous case*; this is the situation in which $d = 0$. We solve the equation (1) by a process of organized guessing: any solution of (1) will be a function that cancels with its derivatives. Thus it is a function that is similar in form to its derivatives. Certainly exponentials fit this description. Thus we guess a solution of the form

$$y = e^{rx}.$$

Plugging this guess into (1) gives

$$a\left[e^{rx}\right]'' + b\left[e^{rx}\right]' + c\left[e^{rx}\right] = 0.$$

Calculating the derivatives, we find that

$$a \cdot r^2 \cdot e^{rx} + b \cdot r \cdot e^{rx} + c \cdot e^{rx} = 0$$

or

$$[ar^2 + br + c] \cdot e^{rx} = 0.$$

This last equation can only be true (for all x) if

$$ar^2 + br + c = 0.$$

Of course this is a simple quadratic equation (called the *associated polynomial equation*), and we may solve it using the quadratic formula. This process will lead to our solution set.

EXAMPLE 2.1

e.g.

Solve the differential equation

$$y'' - 5y' + 4y = 0.$$

SOLUTION

Following the paradigm just outlined, we guess a solution of the form $y = e^{rx}$. This leads to the quadratic equation for r given by

$$r^2 - 5r + 4 = 0.$$

Of course this easily factors into

$$(r - 1)(r - 4) = 0,$$

so $r = 1, 4$.

Thus e^x and e^{4x} are solutions to the differential equation. A *general solution* is given by

$$y = A \cdot e^x + B \cdot e^{4x}, \tag{2}$$

where A and B are arbitrary constants. You may check that any function of the form (2) solves the original differential equation. Observe that our general solution (2) has two undetermined constants, which is consistent with the fact that we are solving a second-order differential equation.

 EXAMPLE 2.2

Solve the differential equation

$$2y'' + 6y' + 2y = 0.$$

SOLUTION

The associated polynomial equation is

$$2r^2 + 6r + 2 = 0.$$

This equation does not factor in any obvious way, so we use the quadratic formula:

$$r = \frac{-6 \pm \sqrt{6^2 - 4 \cdot 2 \cdot 2}}{2 \cdot 2} = \frac{-6 \pm \sqrt{20}}{4} = \frac{-3 \pm \sqrt{5}}{2}.$$

Thus the general solution to the differential equation is

$$y = A \cdot e^{\frac{-3+\sqrt{5}}{2} \cdot x} + B \cdot e^{\frac{-3-\sqrt{5}}{2} \cdot x}.$$

 Math Note: Much of the analysis that we have applied to second-order, constant coefficient, linear equations will apply, virtually without change, to constant coefficient, linear equations of high order. We shall say more about this topic in Section 2.7.

You Try It: Find the general solution of the second-order linear differential
equation

$$y'' - 6y' + 5y = 0.$$

EXAMPLE 2.3

Solve the differential equation

$$y'' - 6y' + 9y = 0.$$

SOLUTION

In this case the associated polynomial is

$$r^2 - 6r + 9 = 0.$$

This algebraic equation has the single solution $r = 3$. But our differential equation is second-order, and therefore we seek *two independent solutions*.

In the case that the associated polynomial has just one root, we find the other solution with an augmented guess: Our new guess is $y = x \cdot e^{3x}$. You may check for yourself that this new guess is also a solution. So the general solution of the differential equation is

$$y = A \cdot e^{3x} + B \cdot xe^{3x}.$$

You Try It: Find the general solution of the differential equation

$$y'' + 4y' + 4y = 0.$$

As a prologue to our next example, we must review some ideas connected with complex exponentials. Recall that, for a real variable x,

$$e^x = 1 + x + \frac{x^2}{2!} + \frac{x^3}{3!} + \cdots = \sum_{j=0}^{\infty} \frac{x^j}{j!}.$$

This equation persists if we replace the real variable x by a complex variable z. Thus

$$e^z = 1 + z + \frac{z^2}{2!} + \frac{z^3}{3!} + \cdots = \sum_{j=0}^{\infty} \frac{z^j}{j!}.$$

Now write $z = x + iy$, and let us gather together the real and imaginary parts of this last equation:

$$e^z = e^{x+iy}$$

$$= e^x \cdot e^{iy}$$

$$= e^x \cdot \left[1 + iy + \frac{(iy)^2}{2!} + \frac{(iy)^3}{3!} + \frac{(iy)^4}{4!} + \cdots \right]$$

$$= e^x \cdot \left[\left(1 - \frac{y^2}{2!} + \frac{y^4}{4!} - + \cdots \right) + i \left(y - \frac{y^3}{3!} + \frac{y^5}{5!} - + \cdots \right) \right]$$

$$= e^x [\cos y + i \sin y].$$

In the special case $x = 0$, the equation

$$e^{iy} = \cos y + i \sin y$$

is known as *Euler's formula*, in honor of Leonhard Euler (1707–1783). We will also make considerable use of the more general formula

$$e^{x+iy} = e^x [\cos y + i \sin y].$$

In using complex numbers, you should of course remember that the square root of a negative number is an *imaginary number*. For instance,

$$\sqrt{-4} = \pm 2i \qquad \text{and} \qquad \sqrt{-25} = \pm 5i.$$

EXAMPLE 2.4

Solve the differential equation

$$4y'' + 4y' + 2y = 0.$$

SOLUTION

The associated polynomial is

$$4r^2 + 4r + 2 = 0.$$

We apply the quadratic equation to solve it:

$$r = \frac{-4 \pm \sqrt{4^2 - 4 \cdot 4 \cdot 2}}{2 \cdot 4} = \frac{-4 \pm \sqrt{-16}}{8} = \frac{-1 \pm i}{2}.$$

Thus the solutions to our differential equation are

$$y = e^{\frac{-1+i}{2} \cdot x} \qquad \text{and} \qquad y = e^{\frac{-1-i}{2} \cdot x}.$$

A general solution is given by

$$y = A \cdot e^{\frac{-1+i}{2} \cdot x} + B \cdot e^{\frac{-1-i}{2} \cdot x}.$$

Using Euler's formula, we may rewrite this general solution as

$$y = A \cdot e^{-x/2}[\cos x/2 + i \sin x/2]$$
$$+ Be^{-x/2}[\cos x/2 - i \sin x/2]. \tag{3}$$

We shall now use some propitious choices of A and B to extract meaningful real-valued solutions. First choose $A = 1/2$, $B = 1/2$. Putting these values in equation (3) gives

$$y = e^{-x/2} \cos x/2.$$

Now taking $A = -i/2$, $B = i/2$ gives the solution

$$y = e^{-x/2} \sin x/2.$$

As a result of this little trick, we may rewrite the general solution to our differential equation as

$$y = A \cdot e^{-x/2} \cos x/2 + B \cdot e^{-x/2} \sin x/2.$$

You Try It: Find the general solution of the differential equation

$$y'' + y' + y = 0.$$

Write this solution without using complex numbers (but certainly use complex numbers to *find* the solution).

Math Note: Complex numbers and complex analysis have a long history. For a long time these numbers were considered to be suspect—they did not really exist, but they had certain uses that made them tolerable. Today we know how to construct the complex numbers in a concrete manner (see [KRA2], [KRA4]).

We conclude this section with a last example of homogeneous, second-order, linear ordinary differential equation with constant coefficients, and with complex roots, just to show how straightforward the methodology really is.

EXAMPLE 2.5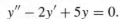
Solve the differential equation

$$y'' - 2y' + 5y = 0.$$

SOLUTION

The associated polynomial is

$$r^2 - 2r + 5 = 0.$$

According to the quadratic formula, the solutions of this equation are

$$r = \frac{2 \pm \sqrt{(-2)^2 - 4 \cdot 1 \cdot 5}}{2} = \frac{2 \pm 4i}{2} = 1 \pm 2i.$$

Hence the roots of the associated polynomial are $r = 1 + 2i$ and $1 - 2i$.

According to what we have learned, two independent solutions to the differential equation are thus given by

$$y = e^x \cos 2x \qquad \text{and} \qquad y = e^x \sin 2x.$$

Therefore the general solution is given by

$$y = A e^x \cos 2x + B e^x \sin 2x.$$

☞ **You Try It:**   Find the general solution of the differential equation

$$2y'' - 3y' + 6y = 0.$$

2.2 The Method of Undetermined Coefficients

"Undetermined coefficients" is a method of organized guessing. We have already seen guessing, in one form or another, serve us well in solving first-order linear equations and also in solving homogeneous second-order linear equations with constant coefficients. Now we shall expand the technique to cover *inhomogeneous* second-order linear equations.

We must begin by discussing what the solution to such an equation will look like. Consider an equation of the form

$$ay'' + by' + cy = f(x). \tag{1}$$

Suppose that we can find (by guessing or by some other means) a function $y = y_0(x)$ that satisfies this equation. We call y_0 a *particular solution* of the differential equation. Notice that it will *not* be the case that a constant multiple of y will also solve the equation. In fact, if we consider $y = A \cdot y_0$ and plug this function into the equation, we obtain

$$a[Ay_0]'' + b[Ay_0]' + c[Ay_0] = A[ay_0'' + by_0' + cy_0] = A \cdot f.$$

But we expect the solution of a second-order equation to have two free constants. Where will they come from?

The answer is that we must separately solve the associated *homogeneous equation*, which is

$$ay'' + by' + cy = 0.$$

If y_1 and y_2 are solutions of this equation, then of course (as we learned in the last section) we know that $A \cdot y_1 + B \cdot y_2$ will be a general solution of *this homogeneous equation*. But then the general solution of the original differential equation (1) will be

$$y = y_0 + A \cdot y_1 + B \cdot y_2.$$

Math Note: We invite you to verify that, no matter what the choice of A and B, this y will be a solution of the original differential equation (1).

These ideas are best hammered home by the examination of some examples.

EXAMPLE 2.6

Find the general solution of the differential equation

$$y'' + y = \sin x. \tag{2}$$

SOLUTION

We might guess that $y = \sin x$ or $y = \cos x$ is a particular solution of this equation. But in fact these are solutions of the homogeneous equation

$$y'' + y = 0$$

(as we may check by using the techniques of the last section, or just by direct verification). So if we want to find a particular solution of (2), then we must try a bit harder.

Inspired by our experience with the case of repeated roots for the second-order, homogeneous linear equation with constant coefficients (as in the last section), we shall instead guess

$$y_0 = \alpha \cdot x \cos x + \beta \cdot x \sin x$$

for our particular solution. Notice that we allow arbitrary constants in front of the functions $x \cos x$ and $x \sin x$. These are the "undetermined coefficients" that we seek.

Now we simply plug the guess into the differential equation and see what happens. Thus

$$[\alpha \cdot x \cos x + \beta \cdot x \sin x]'' + [\alpha \cdot x \cos x + \beta \cdot x \sin x] = 0$$

or

$$\alpha\,(2(-\sin x) + x(-\cos x)) + \beta(2 \cos x$$
$$+ x(-\sin x)) + \alpha x \cos x + \beta x \sin x = 0$$

or

$$(-2\alpha) \sin x + (2\beta) \cos x + (-\beta + \beta)x \sin x + (-\alpha + \alpha)x \cos x = \sin x.$$

We see that there is considerable cancellation, and we end up with

$$-2\alpha \sin x + 2\beta \cos x = \sin x.$$

The only way that this can be an identity in x is if $-2\alpha = 1$ and $2\beta = 0$ or $\alpha = -1/2$ and $\beta = 0$.

Thus our particular solution is

$$y_0 = -\tfrac{1}{2}x \cos x$$

and our general solution is

$$y = -\tfrac{1}{2}x \cos x + A \cos x + B \sin x.$$

EXAMPLE 2.7

Find the solution of

$$y'' - y' - 2y = 4x^2$$

that satisfies $y(0) = 0$ and $y'(0) = 1$.

SOLUTION

The associated homogeneous equation is

$$y'' - y' - 2y = 0$$

and this has associated polynomial

$$r^2 - r - 2 = 0.$$

The roots are obviously $r = 2, -1$ and so the general solution of the associated homogeneous equation is $y = A \cdot e^{2x} + B \cdot e^{-x}$.

For a particular solution, our guess will be a polynomial. Guessing a second-degree polynomial makes good sense, since a guess of a higher-order polynomial is going to produce terms of high degree that we do not want. Thus we guess that $y_p(x) = \alpha x^2 + \beta x + \gamma$. Plugging this guess into the differential equation gives

$$[\alpha x^2 + \beta x + \gamma]'' - [\alpha x^2 + \beta x + \gamma]' - 2[\alpha x^2 + \beta x + \gamma] = 4x^2$$

or

$$[2\alpha] - [\alpha \cdot 2x + \beta] - [2\alpha x^2 + 2\beta x + 2\gamma] = 4x^2.$$

Grouping like terms together gives

$$-2\alpha x^2 + [-2\alpha - 2\beta]x + [2\alpha - \beta - 2\gamma] = 4x^2.$$

As a result, we find that

$$-2\alpha = 4$$

$$-2\alpha - 2\beta = 0$$

$$2\alpha - \beta - 2\gamma = 0.$$

This system is easily solved to yield $\alpha = -2$, $\beta = 2$, $\gamma = -3$. So our particular solution is $y_0(x) = -2x^2 + 2x - 3$. The general solution of the original differential equation is then

$$y(x) = (-2x^2 + 2x - 3) + A \cdot e^{2x} + B \cdot e^{-x}.$$

Now we seek the solution that satisfies the initial conditions $y(0) = 0$ and $y'(0) = 1$. These translate to

$$0 = y(0) = -2 \cdot 0^2 + 2 \cdot 0 - 3 + A \cdot e^0 + B \cdot e^0$$

and

$$1 = y'(0) = -4 \cdot 0 + 2 - 0 + 2A \cdot e^0 - B \cdot e^0.$$

This gives the equations

$$0 = -3 + A + B$$

$$1 = 2 + 2A - B.$$

Of course we can solve this system quickly to find that $A = 1/3$, $B = 8/3$.

In conclusion, the solution to our initial boundary value problem is

$$y(x) = -2x^2 + 2x - 3 + \tfrac{1}{3} \cdot e^{2x} - \tfrac{8}{3} \cdot e^{-x}.$$

You Try It: Solve the differential equation

$$y'' - y = \cos x.$$

You Try It: Find the solution to the initial value problem

$$y'' + y' = x, \quad y(0) = 1, \quad y'(0) = 0.$$

Math Note: If we wish to use the method of undetermined coefficients to solve the differential equation

$$y^{(iv)} + 2y^{(ii)} + y = \sin x,$$

then we must note that $\sin x$, $\cos x$, $x \sin x$, and $x \cos x$ are all solutions of the associated homogeneous equation

$$y^{(iv)} + 2y^{(ii)} + y = 0.$$

Thus we will need to guess a particular solution of the form $Ax^2 \cos x + Bx^2 \sin x$. We invite the reader to try this guess and find a particular solution.

2.3 The Method of Variation of Parameters

Variation of parameters is a method for producing a *particular solution* to an inhomogeneous equation by exploiting the (usually much simpler to find) solutions to the associated homogeneous equation.

Let us consider the differential equation

$$y'' + p(x)y' + q(x)y = r(x). \tag{1}$$

Assume that, by some method or other, we have found the general solution of the associated homogeneous equation

$$y'' + p(x)y' + q(x)y = 0$$

to be

$$y = Ay_1(x) + By_2(x).$$

What we do now is to *guess* that a particular solution to the original equation (1) has the form

$$y_0(x) = v_1(x) \cdot y_1(x) + v_2(x) \cdot y_2(x). \tag{2}$$

Now let us proceed on this guess. We calculate that

$$y_0' = [v_1' y_1 + v_1 y_1'] + [v_2' y_1 + v_2 y_2'] = [v_1' y_1 + v_2' y_2] + [v_1 y_1' + v_2 y_2']. \quad (3)$$

Now we also need to calculate the second derivative of y_0. But we do not want the extra complication of having second derivatives of v_1 and v_2. So we will mandate that the first expression in brackets on the far right side of (3) is identically zero. Thus we have

$$v_1' y_1 + v_2' y_2 \equiv 0. \quad (4)$$

Thus

$$y_0' = v_1 y_1' + v_2 y_2'$$

and we can now calculate that

$$y_0'' = [v_1' y_1' + v_1 y_1''] + [v_2' y_2' + v_2 y_2'']. \quad (5)$$

Now let us substitute (2), (3), and (5) into the differential equation. The result is

$$\left([v_1' y_1' + v_1 y_1''] + [v_2' y_2' + v_2 y_2'']\right) + p(x) \cdot \left(v_1 y_1' + v_2 y_2'\right)$$
$$+ q(z) \cdot (v_1 y_1 + v_2 y_2).$$

After some algebraic manipulation, this becomes

$$v_1 \left(y_1'' + p y_1' + q y_1\right) + v_2 \left(y_2'' + p y_2' + q y_2\right) + v_1' y_1' + v_2' y_2' = r.$$

Since y_1, y_2 are solutions of the homogeneous equation, the expressions in parentheses vanish. The result is

$$v_1' y_1' + v_2' y_2' = r. \quad (6)$$

At long last we have two equations to solve in order to determine what v_1 and v_2 must be. Namely, we use equations (4) and (6) to obtain

$$v_1' y_1 + v_2' y_2 = 0,$$
$$v_1' y_1' + v_2' y_2' = r.$$

In practice, these can be often solved for v_1', v_2', and then integration tells us what v_1, v_2 must be.

As usual, the best way to understand a new technique is by way of some examples.

EXAMPLE 2.8

Find the general solution of

$$y'' + y = \csc x.$$

SOLUTION

Of course the general solution to the associated homogeneous equation is familiar. It is

$$y(x) = A \sin x + B \cos x.$$

In order to find a particular solution, we need to solve the equations

$$v_1' \sin x + v_2' \cos x = 0,$$

$$v_1'(\cos x) + v_2'(-\sin x) = \csc x.$$

This is a simple algebra problem, and we find that

$$v_1'(x) = \cot x \qquad \text{and} \qquad v_2'(x) = -1.$$

As a result,

$$v_1(x) = \ln(\sin x) \qquad \text{and} \qquad v_2(x) = -x.$$

[As you will see, we do not need any constants of integration.]

The final result is then that a particular solution of our differential equation is

$$y_0(x) = v_1(x)y_1(x) + v_2(x)y_2(x) = [\ln(\sin x)] \cdot \sin x + [-x] \cdot \cos x.$$

We invite you to check that this solution actually works. The general solution of the original differential equation is

$$y(x) = ([\ln(\sin x)] \cdot \sin x + [-x] \cdot \cos x) + A \sin x + B \cos x.$$

 EXAMPLE 2.9

Solve the differential equation

$$y'' - y' - 2y = 4x^2$$

using the method of variation of parameters.

SOLUTION

You will note that, in the last section (Example 2.7), we solved this same equation using the method of undetermined coefficients (or organized guessing). Now we will solve it a second time by our new method.

As we saw before, the homogeneous equation has the general solution

$$y = Ae^{2x} + Be^{-x}.$$

Now we solve the system

$$v_1' e^{2x} + v_2' e^{-x} = 0,$$

$$v_1'[2e^{2x}] + v_2'[-e^{-x}] = 4x^2.$$

The result is

$$v_1'(x) = \frac{4}{3}x^2 e^{-2x} \qquad \text{and} \qquad v_2'(x) = -\frac{4}{3}x^2 e^x.$$

We may use integration by parts to then determine that

$$v_1(x) = -\frac{2x^2}{3}e^{-2x} - \frac{2x}{3}e^{-2x} - \frac{1}{3}e^{-2x}$$

and

$$v_2(x) = -\frac{4x^2}{3}e^x + \frac{8x}{3}e^x - \frac{8}{3}e^x.$$

We finally see that a particular solution to our differential equation is

$$y_0(x) = v_1(x) \cdot y_1(x) + v_2(x)y_2(x)$$

$$= \left[-\frac{2x^2}{3}e^{-2x} - \frac{2x}{3}e^{-2x} - \frac{1}{3}e^{-2x}\right] \cdot e^{2x}$$

$$+ \left[-\frac{4x^2}{3}e^x + \frac{8x}{3}e^x - \frac{8}{3}e^x\right] \cdot e^{-x}$$

$$= \left[-\frac{2x^2}{3} - \frac{2x}{3} - \frac{1}{3}\right] + \left[-\frac{4x^2}{3} + \frac{8x}{3} - \frac{8}{3}\right]$$

$$= -2x^2 + 2x - 3.$$

In conclusion, the general solution of the original differential equation is

$$y(x) = \left(-2x^2 + 2x - 3\right) + Ae^{2x} + Be^{-x}.$$

As you can see, this is the same answer that we obtained in Section 2.2, Example 2.7, by the method of undetermined coefficients.

You Try It: Use the method of variation of parameters to find the general solution of the differential equation

$$y'' - 2y = x + 1.$$

You Try It: Use the method of this section to solve the initial value problem

$$y'' + 3y' + 2y = \cos x, \quad y(0) = 0, \quad y'(0) = 2.$$

Math Note: Notice that the method of variation of parameters *always* gives a system of two equations in two unknowns that we can solve for v_1' and v_2'. After that, it might be tricky to solve for v_1 and v_2. Even so, it can be useful to know v_1' and v_2'. Numerical integration techniques and other devices can still be used to obtain information about the solution of the original differential equation.

2.4 The Use of a Known Solution to Find Another

Consider a general second-order linear equation of the form

$$y'' + p(x)y' + q(x)y = 0. \tag{1}$$

It often happens—and we have seen this in our earlier work—that one can either guess or elicit one solution to the equation. But finding the second independent solution is more difficult. In this section we introduce a method for finding that second solution.

In fact we exploit a notational trick that served us well in Section 2.3 on variation of parameters. Namely, we will assume that we have found the one solution y_1 and we will suppose that the second solution we seek is $y_2 = v \cdot y_1$ for some undetermined factor v. Our job, then, is to find v.

Assuming, then, that y_1 is a solution of (1), we will substitute $y_2 = v \cdot y_1$ into (1) and see what this tells us about calculating v. We see that

$$[v \cdot y_1]'' + p(x) \cdot [v \cdot y_1]' + r(x) \cdot [v \cdot y_1] = 0$$

or

$$[v'' \cdot y_1 + 2v' \cdot y_1' + v \cdot y_1''] + p(x) \cdot [v' \cdot y_1 + v \cdot y_1'] + r(x) \cdot [v \cdot y_1] = 0.$$

We rearrange this identity to find that

$$v \cdot [y_1'' + p(x) \cdot y_1' + y_1] + [v'' \cdot y_1] + [v' \cdot (2y_1' + p \cdot y_1)] = 0.$$

Now we are *assuming* that y_1 is a solution of the differential equation (1), so the first expression in brackets must vanish. As a result,

$$[v'' \cdot y_1] + [v' \cdot (2y_1' + p \cdot y_1)] = 0.$$

In the spirit of separation of variables, we may rewrite this equation as

$$\frac{v''}{v'} = -2\frac{y_1'}{y_1} - p.$$

Integrating once, we find that

$$\ln v' = -2\ln y_1 - \int p(x)\,dx$$

or

$$v' = \frac{1}{y_1^2}e^{-\int p(x)\,dx}.$$

Applying the integral one last time yields

$$v = \int \frac{1}{y_1^2}e^{-\int p(x)\,dx}\,dx. \qquad (2)$$

In order to really understand what this means, let us apply the method to some particular differential equations.

EXAMPLE 2.10
Find the general solution of the differential equation

$$y'' - 4y' + 4y = 0.$$

SOLUTION
When we first encountered this type of equation in Section 2.1, we learned to study the associated polynomial

$$r^2 - 4y + 4 = 0.$$

Unfortunately, the polynomial has only the repeated root $r = 2$, so we at first find just the one solution $y_1(x) = e^{2x}$. Where do we find another?

In Section 2.1, we found the second solution by guessing. Now we have a more systematic way of finding that second solution, and we use it now to test out our new methodology. Observe that $p(x) = -4$ and $q(x) = 4$.

According to formula (2), we can find a second solution $y_2 = v \cdot y_1$ with

$$v = \int \frac{1}{y_1^2} e^{-\int p(x)\,dx}\,dx$$

$$= \int \frac{1}{[e^{2x}]^2} e^{-\int -4\,dx}\,dx$$

$$= \int e^{-4x} \cdot e^{4x}\,dx$$

$$= \int 1\,dx = x.$$

Thus the second solution to our differential equation is $y_2 = v \cdot y_1 = x \cdot e^{2x}$ and the general solution is

$$y = A \cdot e^{2x} + B \cdot xe^{2x}.$$

Next we turn to an example of a nonconstant coefficient equation.

EXAMPLE 2.11

Find the general solution of the differential equation

$$x^2 y'' + xy' - y = 0.$$

SOLUTION

Differentiating a monomial once lowers the degree by 1 and differentiating it twice lowers the degree by 2. So it is natural to guess that this differential equation has a power of x as a solution. And $y_1(x) = x$ works.

We use formula (2) to find a second solution of the form $y_2 = v \cdot y_1$. First we rewrite the equation in the standard form as

$$y'' + \frac{1}{x}y' - \frac{1}{x^2}y = 0$$

and we note then that $p(x) = 1/x$ and $q(x) = -1/x^2$. Thus we see that

$$v(x) = \int \frac{1}{y_1^2} e^{-\int p(x)\,dx}\,dx$$

$$= \int \frac{1}{x^2} e^{-\int 1/x\,dx}\,dx$$

$$= \int \frac{1}{x^2} e^{-\ln x}\,dx$$

$$= \int \frac{1}{x^2} \frac{1}{x} \, dx$$

$$= -\frac{1}{2x^2}.$$

In conclusion, $y_2 = v \cdot y_1 = [-1/(2x^2)] \cdot x = -1/(2x)$ and the general solution is

$$y(x) = A \cdot x + B \cdot \left(-\frac{1}{2x}\right).$$

You Try It: Use the methodology of this section to find the general solution of   the differential equation

$$y'' - \frac{1}{2x} y' - \frac{1}{x^2} y = 0.$$

[*Hint:* One solution will be a positive, integer power of x.]

Math Note: As with the method of variation of parameters, we find with this new technique for finding a second solution that we will *always* be able to write down the integral for v. Whether we will actually be able to evaluate the integral and find v explicitly will depend on the particular problem that we are studying. But, even when the integral cannot be explicitly evaluated, we can use numerical and other techniques to obtain information about v and then about y_2.

2.5 Vibrations and Oscillations

When a physical system in stable equilibrium is disturbed, then it is subject to forces that tend to restore the equilibrium. The result is a system that can lead to oscillations or vibrations. It is described by an ordinary differential equation of the form

$$\frac{d^2x}{dt^2} + p(t) \cdot \frac{dx}{dt} + q(t)x = r(t).$$

In this section we shall learn how and why such an equation models the physical system we have described, and we shall see how its solution sheds light on the physics of the situation.

2.5.1 UNDAMPED SIMPLE HARMONIC MOTION

Our basic example will be a cart of mass M attached to a nearby wall by means of a spring. See Fig. 2.1. The spring exerts no force when the cart is at its rest position $x = 0$. According to Hooke's Law, if the cart is displaced a distance x, then the spring exerts a proportional force $F_s = -kx$, where k is a positive constant known as Hooke's constant. Observe that if $x > 0$ then the cart is moved to the right and the spring pulls to the left; so the force is negative. Obversely, if $x < 0$ then the cart is moved to the left and the spring resists with a force to the right; so the force is positive.

Newton's second law of motion says that the mass of the cart times its acceleration equals the force acting on the cart. Thus

$$M \cdot \frac{d^2 x}{dt^2} = F_s = -k \cdot x.$$

As a result,

$$\frac{d^2 x}{dt^2} + \frac{k}{M} x = 0.$$

It is both convenient and traditional to let $a = \sqrt{k/M}$ (both k and M are positive) and thus to write the equation as

$$\frac{d^2 x}{dt^2} + a^2 x = 0.$$

Of course this is a familiar differential equation for us, and we can write its general solution immediately:

$$x(t) = A \sin at + B \cos at.$$

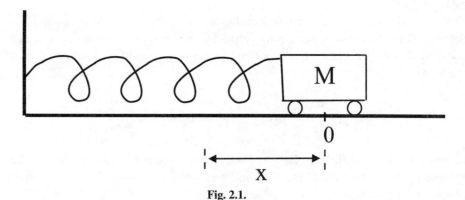

Fig. 2.1.

Now suppose that the cart is pulled to the right to an initial position of $x = x_0$ and then is simply released (with initial velocity 0). Then we have the initial conditions

$$x(0) = x_0 \quad \text{and} \quad \frac{dx}{dt}(0) = 0.$$

Thus

$$x_0 = A \sin(a \cdot 0) + B \cos(a \cdot 0)$$

$$0 = Aa \cos(a \cdot 0) - Ba \sin(a \cdot 0)$$

or

$$x_0 = B$$

$$0 = A \cdot a.$$

We conclude that $B = x_0$, $A = 0$, and we find the solution of the system to be

$$x(t) = x_0 \cos at.$$

In other words, if the cart is displaced a distance x_0 and released, then the result is a simple harmonic motion (described by the cosine function) with *amplitude* x_0 (i.e., the cart glides back and forth, x_0 units to the left of the origin and then x_0 units to the right) and with period $T = 2\pi/a$ (which means that the motion repeats itself every $2\pi/a$ units of time).

The *frequency* f of the motion is the number of cycles per unit of time, hence $f \cdot T = 1$, or $f = 1/T = a/(2\pi)$. It is useful to substitute back in the actual value of a so that we can analyze the physics of the system. Thus

$$\text{Amplitude} = x_0$$

$$\text{Period} = T = \frac{2\pi \sqrt{M}}{\sqrt{k}}$$

$$\text{Frequency} = f = \frac{\sqrt{k}}{2\pi \sqrt{M}}.$$

Math Note: We see that if the stiffness of the spring k is increased then the period becomes smaller and the frequency increases. Likewise, if the mass of the cart is increased then the period increases and the frequency decreases. Thus the mathematical analysis coincides with, and reinforces, our physical intuition.

2.5.2 DAMPED VIBRATIONS

It probably has occurred to you that the physical model in the last subsection is not realistic. Typically, a cart that is attached to a spring and released, just as we have described, will enter a harmonic motion that *dies out over time*. In other words, resistance and friction will cause the system to be damped. Let us add that information to the system.

Physical considerations make it plausible to postulate that the resistance is proportional to the velocity of the moving cart. Thus the resistive force is

$$F_d = -c\frac{dx}{dt},$$

where F_d denotes damping force and $c > 0$ is a positive constant that measures the resistance of the medium (air or water or oil, etc.). Notice, therefore, that when the cart is traveling to the right, then $dx/dt > 0$ and therefore the force of resistance is negative (i.e., in the other direction). Likewise, when the cart is traveling to the left, then $dx/dt < 0$ and the force of resistance is positive.

Since the total of all the forces acting on the cart equals the mass times the acceleration, we now have

$$M \cdot \frac{d^2x}{dt^2} = F_s + F_d.$$

In other words,

$$\frac{d^2x}{dt^2} + \frac{c}{M} \cdot \frac{dx}{dt} + \frac{k}{M} \cdot x = 0.$$

Because of convenience and tradition, we again take $a = \sqrt{k/M}$ and we set $b = c/(2M)$. Thus the differential equation takes the form

$$\frac{d^2x}{dt^2} + 2b \cdot \frac{dx}{dt} + a^2 \cdot x = 0.$$

This is a second-order, linear, homogeneous ordinary differential equation with constant coefficients. The associated polynomial is

$$r^2 + 2br + a^2 = 0,$$

and it has roots

$$r_1, r_2 = \frac{-2b \pm \sqrt{4b^2 - 4a^2}}{2} = -b \pm \sqrt{b^2 - a^2}.$$

Now we must consider three cases.

Case A: $b^2 - a^2 > 0$ In words, we are assuming that the frictional force (which depends on c) is significantly larger than the stiffness of the spring (which depends on k). Thus we would expect the system to damp heavily. In any event, the calculation of r_1, r_2 involves the square root of a *positive real number*, and thus r_1, r_2 are distinct real (and negative) roots of the associated polynomial equation.

Thus the general solution of our system in this case is

$$x = Ae^{r_1 t} + Be^{r_2 t},$$

where (we repeat) r_1, r_2 are negative real numbers. We apply the initial conditions $x(0) = x_0$, $dx/dt(0) = 0$, just as in the last section (details are left to you). The result is the particular solution

$$x(t) = \frac{x_0}{r_1 - r_2} \left(r_1 e^{r_2 t} - r_2 e^{r_1 t} \right). \tag{1}$$

Notice that, in this heavily damped system, no oscillation occurs (i.e., there are no sines or cosines in the expression for $x(t)$). The system simply dies out. Figure 2.2 exhibits the graph of the function in (1).

Math Note: The type of harmonic motion illustrated in this last discussion, and in Fig. 2.2, is the ideal motion that is induced by the resistance of a shock absorber on an automobile. The whole purpose of a shock absorber is to make the harmonic motion, that would be induced by the car hitting a bump, die out immediately.

Case B: $b^2 - a^2 = 0$ This is the critical case, where the resistance balances the force of the spring. We see that $b = a$ (both are known to be positive) and $r_1 = r_2 = -b = -a$. We know, then, that the general solution to our differential

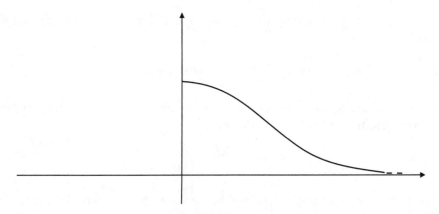

Fig. 2.2.

equation is

$$x(t) = Ae^{-at} + Bte^{-at}.$$

When the standard initial conditions are imposed, we find the particular solution

$$x(t) = x_0 \cdot e^{-at}(1 + at).$$

We see that this differs from the situation in Case A by the factor $(1 + at)$. That factor of course attenuates the damping, but there is *still no oscillatory motion*. We call this the *critical case*. The graph of our new $x(t)$ is quite similar to the graph already shown in Fig. 2.2.

Math Note: When a shock absorber begins to wear out, it becomes less effective. At a certain critical stage its action will be described more accurately by Case B than by Case A. In physical terms, this will mean that the oscillations of the car (induced by a road bump, for instance) will be damped out less effectively. The car will return to true more slowly.

If there is any small decrease in the viscosity, however slight, then the system will begin to vibrate (as one would expect). That is the next, and last, case that we examine.

Case C: $b^2 - a^2 < 0$ Now $0 < b < a$ and the damping is less than the force of the spring. The calculation of r_1, r_2 entails taking the square root of a negative number. Thus r_1, r_2 are the conjugate complex numbers $-b \pm i\sqrt{a^2 - b^2}$. We set $\alpha = \sqrt{a^2 - b^2}$.

Now the general solution of our system, as we well know, is

$$x(t) = e^{-bt}(A \sin \alpha t + B \cos \alpha t).$$

If we evaluate A, B according to our usual initial conditions, then we find the particular solution

$$x(t) = \frac{x_0}{\alpha} e^{-bt}(b \sin \alpha t + \alpha \cos \alpha t).$$

It is traditional and convenient to set $\theta = \tan^{-1}(b/\alpha)$. With this notation, we can express the last equation in the form

$$x(t) = \frac{x_0\sqrt{\alpha^2 + b^2}}{\alpha} e^{-bt} \cos(\alpha t - \theta). \tag{2}$$

As you can see, there is oscillation because of the presence of the cosine function. The amplitude (the expression that appears in front of cosine) clearly

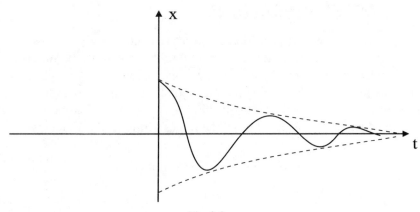

Fig. 2.3.

falls off—rather rapidly—with t because of the presence of the exponential. The graph of this function is exhibited in Fig. 2.3.

Of course this function is *not* periodic—it is dying off, and not repeating itself. What *is true*, however, is that the graph crosses the t-axis (the equilibrium position $x = 0$) at regular intervals. If we consider this interval T (which is not a "period," strictly speaking) as the time required for one complete cycle, then $\alpha T = 2\pi$, so

$$T = \frac{2\pi}{\alpha} = \frac{2\pi}{\sqrt{k/M - c^2/(4M^2)}}. \tag{3}$$

We define the number f, which plays the role of "frequency" with respect to the indicated time interval, to be

$$f = \frac{1}{T} = \frac{1}{2\pi}\sqrt{\frac{k}{M} - \frac{c^2}{4M^2}}.$$

This number is commonly called the *natural frequency* of the system. When the viscosity vanishes, then our solution clearly reduces to the one we found earlier when there was no viscosity present. We also see that the frequency of the vibration is reduced by the presence of damping; increasing the viscosity further reduces the frequency.

Math Note: When the shock absorbers on your car are really shot, then the motion of the car after striking a bump will resemble that shown in Fig. 2.3. The harmonic motion begun by the bump will die out—but rather slowly. The result is discomfort for the passengers.

2.5.3 FORCED VIBRATIONS

The vibrations that we have considered so far are called *free vibrations* because all the forces acting on the system are internal to the system itself. We now consider the situation in which there is an *external force* $F_e = f(t)$ acting on the system. This force could be an external magnetic field (acting on the steel cart) or vibration of the wall, or perhaps a stiff wind blowing. Again setting mass times acceleration equal to the resultant of all the forces acting on the system, we have

$$M \cdot \frac{d^2x}{dt^2} = F_s + F_d + F_e.$$

Taking into account the definitions of the various forces, we may write the differential equation as

$$M\frac{d^2x}{dt^2} + c\frac{dx}{dt} + kx = f(t).$$

So we see that the equation describing the physical system is second-order linear, and that the external force gives rise to an inhomogeneous term on the right. An interesting special case occurs when $f(t) = F_0 \cdot \cos \omega t$, in other words when that external force is periodic. Thus our equation becomes

$$M\frac{d^2x}{dt^2} + c\frac{dx}{dt} + kx = F_0 \cdot \cos \omega t. \tag{4}$$

If we can find a particular solution of this equation, then we can combine it with the information about the solution of the associated homogeneous equation in the last subsection and then come up with the general solution of the differential equation. We will use the method of undetermined coefficients. Considering the form of the right-hand side, our guess will be

$$x(t) = \alpha \sin \omega t + \beta \cos \omega t.$$

Substituting this guess into the differential equation gives

$$M\frac{d^2}{dt^2}[\alpha \sin \omega t + \beta \cos \omega t] + c\frac{d}{dt}[\alpha \sin \omega t + \beta \cos \omega t]$$

$$+ k[\alpha \sin \omega t + \beta \cos \omega t] = F_0 \cdot \cos \omega t.$$

With a little calculus and a little algebra we are led to the algebraic equations

$$\omega c \alpha + (k - \omega^2 M)\beta = F_0$$

$$(k - \omega^2 M)\alpha - \omega c \beta = 0.$$

We solve for α and β to obtain

$$\alpha = \frac{\omega c F_0}{(k - \omega^2 M)^2 + \omega^2 c^2} \quad \text{and} \quad \beta = \frac{(k - \omega^2 M) F_0}{(k - \omega^2 M)^2 + \omega^2 c^2}.$$

Thus we have found the particular solution

$$x_0(t) = \frac{F_0}{(k - \omega^2 M)^2 + \omega^2 c^2} \left[\omega c \sin \omega t + (k - \omega^2 M) \cos \omega t \right].$$

We may write this in a more useful form with the notation $\phi = \tan^{-1}[\omega c/(k - \omega^2 M)]$. Thus

$$x_0(t) = \frac{F_0}{\sqrt{(k - \omega^2 M)^2 + \omega^2 c^2}} \cdot \cos(\omega t - \phi). \tag{5}$$

If we assume that we are dealing with the underdamped system, which is Case C of the last subsection, we find that the general solution of our differential equation with periodic external forcing term is

$$x(t) = e^{-bt} \left(A \cos \alpha t + B \sin \alpha t \right)$$

$$+ \frac{F_0}{\sqrt{(k - \omega^2 M)^2 + \omega^2 c^2}} \cdot \cos(\omega t - \phi).$$

We see that, as long as some damping is present in the system (that is, b is nonzero and positive), then the first term in the definition of $x(t)$ is clearly transient (i.e., it dies as $t \to \infty$ because of the exponential term). Thus, as time goes on, the motion assumes the character of the second term in $x(t)$, which is the *steady-state* term. So we can say that, for large t, the physical nature of the general solution to our system is more or less like that of the particular solution $x_0(t)$ that we found. The frequency of this forced vibration equals the impressed frequency (originating with the external forcing term) $\omega/2\pi$. The amplitude is the coefficient

$$\frac{F_0}{\sqrt{(k - \omega^2 M)^2 + \omega^2 c^2}}. \tag{6}$$

This expression for the amplitude depends on all the relevant physical constants, and it is enlightening to analyze it a bit. Observe, for instance, that if the viscosity c is very small and if ω is close to $\sqrt{k/M}$ (so that $k - \omega^2 M$ is very small), then the motion is lightly damped and the external (impressed) frequency $\omega/2\pi$ is close to the natural frequency

$$\frac{1}{2\pi} \sqrt{\frac{k}{M} - \frac{c^2}{4M^2}},$$

and the amplitude is very large (because we are dividing by a number close to 0). This phenomenon is known as *resonance*.

Math Note: There are classical examples of resonance. For instance, several years ago there was a celebration of the anniversary of the Golden Gate Bridge (built in 1937), and many thousands of people marched in unison across the bridge. The frequency of their footfalls was so close to the natural frequency of the bridge (thought of as a suspended string under tension) that the bridge nearly fell apart.

2.5.4 A FEW REMARKS ABOUT ELECTRICITY

It is known that if a periodic electromotive force, $E = E_0$, acts in a simple circuit containing a resistor, an inductor, and a capacitor, then the charge Q on the capacitor is governed by the differential equation

$$L \frac{d^2 Q}{dt^2} + R \frac{dQ}{dt} + \frac{1}{C} Q = E_0 \cos \omega t.$$

This equation is of course quite similar to the equation (4) for the oscillating cart with external force. In particular, the following correspondences (or analogies) are suggested:

$$\text{Mass } M \leftrightarrow \text{Inductance } L$$

$$\text{Viscosity } c \leftrightarrow \text{Resistance } R$$

$$\text{Stiffness of spring } k \leftrightarrow \text{Reciprocal of capacitance } \frac{1}{C}$$

$$\text{Displacement } x \leftrightarrow \text{Charge } Q \text{ on capacitor.}$$

The analogy between the mechanical and electrical systems renders identical the mathematical analysis of the two systems, and enables us to carry over at once all mathematical conclusions from the first to the second. In the given electric circuit we therefore have a critical resistance below which the free behavior of the circuit will be vibratory with a certain natural frequency, a forced steady-state vibration of the charge Q, and resonance phenomena that appear when the circumstances are favorable.

2.6 Newton's Law of Gravitation and Kepler's Laws

Newton's Law of Universal Gravitation is one of the great ideas of modern physics. It underlies so many important physical phenomena that it is part of the bedrock of science. In this section we show how Kepler's laws of planetary motion can be derived from Newton's gravitation law. It might be noted that Johannes Kepler himself (1571–1630) used thousands of astronomical observations (made by Tycho Brahe, 1546–1601) in order to formulate his laws. Both Brahe and Kepler were followers of Copernicus, who postulated that the planets orbited about the sun (rather than the traditional notion that the Earth was the center of the orbits); but Copernicus believed that the orbits were circles. Newton determined how to derive the laws of motion analytically, and he was able to *prove* that the orbits must be ellipses. Furthermore, the eccentricity of an elliptical orbit has an important physical interpretation. The present section explores all these ideas.

Kepler's laws of planetary motion

I.   The orbit of each planet is an ellipse with the sun at one focus (Fig. 2.4).
II. The segment from the center of the sun to the center of an orbiting planet sweeps out area at a constant rate (Fig. 2.5).
III. The square of the period of revolution of a planet is proportional to the cube of the length of the major axis of its elliptical orbit, with the same constant of proportionality for any planet (Fig. 2.6).

It turns out that the eccentricities of the ellipses that arise in the orbits of the planets are very small, so that the orbits are nearly circles, but they are definitely *not* circles. That is the importance of Kepler's First Law.

The second law tells us that when the planet is at its apogee (furthest from the sun), then it is traveling relatively slowly whereas at its perigee (nearest point to the sun), it is traveling relatively rapidly—Fig. 2.7.

The third law allows us to calculate the length of a year on any given planet from knowledge of the shape of its orbit.

In this section we shall learn how to derive Kepler's three laws from Newton's inverse square law of gravitational attraction. To keep matters as simple as possible, we shall assume that our solar system contains a fixed sun and just one planet (the Earth for instance). The problem of analyzing the gravitation influence of three or more planets on each other is incredibly complicated and is still not thoroughly understood.

The argument that we present is due to S. Kochen and is used with his permission.

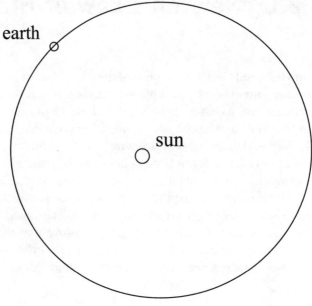

Fig. 2.4.

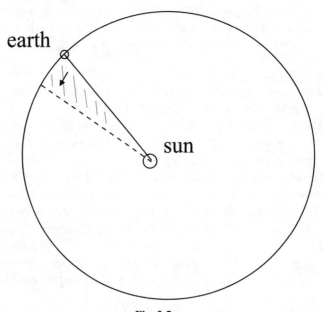

Fig. 2.5.

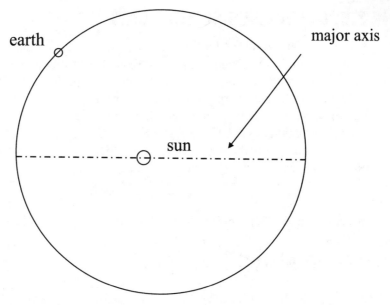

earth

major axis

sun

Fig. 2.6.

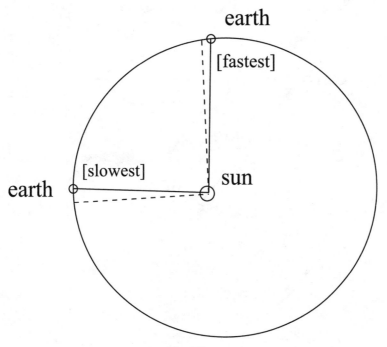

earth

[fastest]

[slowest]

sun

earth

Fig. 2.7.

2.6.1 KEPLER'S SECOND LAW

It is convenient to derive the second law first. We use a polar coordinate system with the origin at the center of the sun. We analyze a single planet which orbits the sun, and we denote the position of that planet at time t by $\mathcal{R}(t)$. The only physical facts that we shall use in this portion of the argument are Newton's second law and the self-evident assertion that the gravitational force exerted by the sun on a planet is a vector parallel to $\mathcal{R}(t)$. See Fig. 2.8.

If $\mathbf{F}$ is force, m is the mass of the planet (Earth), and $\mathbf{a}$ is its acceleration, then Newton's Second Law says that

$$\mathbf{F} = m\mathbf{a} = m\mathcal{R}''(t).$$

We conclude that $\mathcal{R}(t)$ is parallel to $\mathcal{R}''(t)$ for every value of t.

Now

$$\frac{d}{dt}\left(\mathcal{R}(t) \times \mathcal{R}'(t)\right) = \left[\mathcal{R}'(t) \times \mathcal{R}'(t)\right] + \left[\mathcal{R}(t) \times \mathcal{R}''(t)\right].$$

Note that the first of these terms is zero because the cross product of any vector with itself is zero. The second is zero because $\mathcal{R}(t)$ is parallel with $\mathcal{R}''(t)$ for every t. We conclude that

$$\mathcal{R}(t) \times \mathcal{R}'(t) = \mathbf{C}, \tag{1}$$

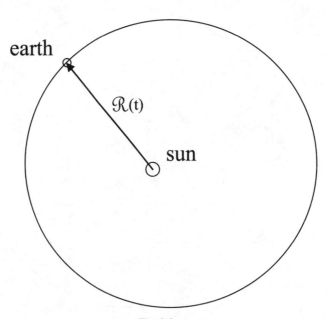

Fig. 2.8.

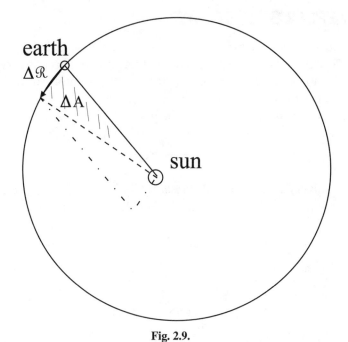

earth

$\Delta \mathcal{R}$

ΔA

sun

Fig. 2.9.

where $\mathbf{C}$ is a constant vector. Notice that this already guarantees that $\mathcal{R}(t)$ and $\mathcal{R}'(t)$ always lie in the same plane, hence that the orbit takes place in a plane.

Now let Δt be an increment of time, $\Delta \mathcal{R}$ the corresponding increment of position, and ΔA the increment of area swept out. Look at Fig. 2.9.

We see that ΔA is approximately equal to half the area of the parallelogram determined by the vectors $\mathcal{R}$ and $\Delta \mathcal{R}$. The area of this parallelogram is $\|\mathcal{R} \times \Delta \mathcal{R}\|$. Thus

$$\frac{\Delta A}{\Delta t} \approx \frac{1}{2} \frac{\|\mathcal{R} \times \Delta \mathcal{R}\|}{\Delta t} = \frac{1}{2} \left\| \mathcal{R} \times \frac{\Delta \mathcal{R}}{\Delta t} \right\|.$$

Letting $\Delta t \to 0$ gives

$$\frac{dA}{dt} = \frac{1}{2} \left\| \mathcal{R} \times \frac{d\mathcal{R}}{dt} \right\| = \frac{1}{2} \|\mathbf{C}\| = \text{constant}.$$

We conclude that area $A(t)$ is swept out at a constant rate. That is Kepler's Second Law.

2.6.2 KEPLER'S FIRST LAW

Now we write $\mathcal{R}(t) = r(t)\mathbf{u}(t)$, where $\mathbf{u}$ is a unit vector pointing in the same direction as $\mathcal{R}$ and r is a positive, scalar-valued function representing the length of $\mathcal{R}$. We use Newton's Inverse Square Law for the attraction of two bodies. If one body (the sun) has mass M and the other (the planet) has mass m, then Newton says that the force exerted by gravity on the planet is

$$-\frac{GmM}{r^2}\mathbf{u}.$$

Here G is a universal gravitational constant. Refer to Fig. 2.10. Because this force is also equal to $m\mathcal{R}''$ (by Newton's Second Law), we conclude that

$$\mathcal{R}'' = -\frac{GM}{r^2}\mathbf{u}.$$

Also

$$\mathcal{R}'(t) = \frac{d}{dt}(r\mathbf{u}) = r'\mathbf{u} + r\mathbf{u}'$$

and

$$0 = \frac{d}{dt}1 = \frac{d}{dt}(\mathbf{u} \cdot \mathbf{u}) = 2\mathbf{u} \cdot \mathbf{u}'.$$

Therefore

$$\mathbf{u} \perp \mathbf{u}'. \tag{2}$$

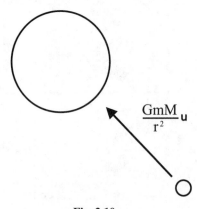

$$\frac{GmM}{r^2}\mathbf{u}$$

Fig. 2.10.

Now, using (1), we calculate

$$\mathcal{R}'' \times \mathbf{C} = \mathcal{R}'' \times (\mathcal{R} \times \mathcal{R}'(t))$$

$$= -\frac{GM}{r^2}\mathbf{u} \times \left(r\mathbf{u} \times (r'\mathbf{u} + r\mathbf{u}')\right)$$

$$= -\frac{GM}{r^2}\mathbf{u} \times (r\mathbf{u} \times r\mathbf{u}')$$

$$= -GM\left[\mathbf{u} \times (\mathbf{u} \times \mathbf{u}')\right].$$

We can determine the vector $\mathbf{u} \times (\mathbf{u} \times \mathbf{u}')$. For, using (2), we see that $\mathbf{u}$ and $\mathbf{u}'$ are perpendicular and that $\mathbf{u} \times \mathbf{u}'$ is perpendicular to both of these. Because $\mathbf{u} \times (\mathbf{u} \times \mathbf{u}')$ is perpendicular to the first and last of these three, it must therefore be parallel to $\mathbf{u}'$. It also has the same length as $\mathbf{u}'$ and, by the right hand rule, points in the opposite direction. Look at Fig. 2.11. We conclude that $\mathbf{u} \times (\mathbf{u} \times \mathbf{u}') = -\mathbf{u}'$, hence that

$$\mathcal{R}'' \times \mathbf{C} = GM\mathbf{u}'.$$

If we antidifferentiate this last equality we obtain

$$\mathcal{R}'(t) \times \mathbf{C} = GM(\mathbf{u} + \mathbf{K}),$$

where $\mathbf{K}$ is a constant vector of integration.

Thus we have

$$\mathcal{R} \cdot (\mathcal{R}'(t) \times \mathbf{C}) = r\mathbf{u}(t) \cdot GM(\mathbf{u}(t) + \mathbf{K}) = GMr(1 + \mathbf{u}(t) \cdot \mathbf{K}),$$

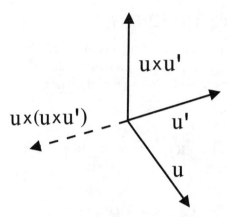

Fig. 2.11.

because $\mathbf{u}(t)$ is a unit vector. If $\theta(t)$ is the angle between $\mathbf{u}(t)$ and $\mathbf{K}$, then we may rewrite our equality as

$$\mathcal{R} \cdot (\mathcal{R}' \times \mathbf{C}) = GMr(1 + \|\mathbf{K}\| \cos \theta).$$

By a standard triple product formula,

$$\mathcal{R} \cdot (\mathcal{R}'(t) \times \mathbf{C}) = (\mathcal{R} \times \mathcal{R}'(t)) \cdot \mathbf{C},$$

which in turn equals

$$\mathbf{C} \cdot \mathbf{C} = \|\mathbf{C}\|^2.$$

[Here we have used the fact, which we derived in the proof of Kepler's Second Law, that $\mathcal{R} \times \mathcal{R}' = \mathbf{C}$.]

Thus

$$\|\mathbf{C}\|^2 = GMr(1 + \|\mathbf{K}\| \cos \theta).$$

[Notice that this equation can be true only if $\|\mathbf{K}\| \le 1$. This fact will come up again below.]

We conclude that

$$r = \frac{\|\mathbf{C}\|^2}{GM} \cdot \left(\frac{1}{1 + \|\mathbf{K}\| \cos \theta} \right).$$

This is the polar equation for an ellipse of eccentricity $\|\mathbf{K}\|$. [Exercise 11 will say a bit more about the such polar equations.]

We have verified Kepler's First Law.

2.6.3 KEPLER'S THIRD LAW

Look at Fig. 2.12. The length $2a$ of the major axis of our elliptical orbit is equal to the maximum value of r plus the minimum value of r. From the equation for the ellipse we see that these occur respectively when $\cos \theta$ is $+1$ and when $\cos \theta$ is -1. Thus

$$2a = \frac{\|\mathbf{C}\|^2}{GM} \frac{1}{1 - \|\mathbf{K}\|} + \frac{\|\mathbf{C}\|^2}{GM} \frac{1}{1 + \|\mathbf{K}\|} = \frac{2\|\mathbf{C}\|^2}{GM(1 - \|\mathbf{K}\|^2)}.$$

We conclude that

$$\|\mathbf{C}\| = \left[aGM(1 - \|\mathbf{K}\|^2) \right]^{1/2}. \tag{3}$$

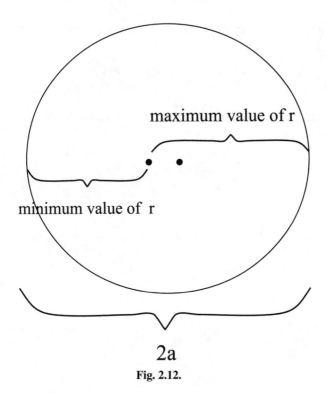

maximum value of r

minimum value of r

2a

Fig. 2.12.

Now recall from our proof of the Second Law that

$$\frac{dA}{dt} = \frac{1}{2}\|\mathbf{C}\|.$$

Then, by antidifferentiating, we find that

$$A(t) = \frac{1}{2}\|\mathbf{C}\|t.$$

(There is no constant term since $A(0) = 0$.) Let $\mathcal{A}$ denote the total area inside the elliptical orbit and T the time it takes to sweep out one orbit. Then

$$\mathcal{A} = A(T) = \frac{1}{2}\|\mathbf{C}\|T.$$

Solving for T we obtain

$$T = \frac{2\mathcal{A}}{\|\mathbf{C}\|}.$$

But the area inside an ellipse with major axis $2a$ and minor axis $2b$ is

$$\mathcal{A} = \pi ab = \pi a^2(1 - e^2)^{1/2},$$

where e is the eccentricity of the ellipse. This equals $\pi a^2(1 - \|\mathbf{K}\|^2)^{1/2}$ by Kepler's First Law. Therefore

$$T = \frac{2\pi a^2(1 - \|\mathbf{K}\|^2)^{1/2}}{\|\mathbf{C}\|}.$$

Finally, we may substitute (3) into this last equation to obtain

$$T = \frac{2\pi a^{3/2}}{(GM)^{1/2}}$$

or

$$\frac{T^2}{a^3} = \frac{4\pi^2}{GM}.$$

This is Kepler's Third Law.

 EXAMPLE 2.12
The planet Uranus describes an elliptical orbit about the sun. It is known that the semi-major axis of this orbit has length 2870×10^6 kilometers. The gravitational constant is $G = 6.637 \times 10^{-8}$ cm^3/(g $\cdot$ sec^2). Finally, the mass of the sun is 2×10^{33} grams. Determine the period of the orbit of Uranus.

SOLUTION
Refer to the explicit formulation of Kepler's Third Law that we proved above. We have

$$\frac{T^2}{a^3} = \frac{4\pi^2}{GM}.$$

We must be careful to use consistent units. The gravitational constant G is given in terms of grams, centimeters, and seconds. The mass of the sun is in grams. We convert the semi-major axis to centimeters: $a = 2870 \times 10^{11}$ cm $= 2.87 \times 10^{14}$ cm. Then we calculate that

$$T = \left(\frac{4\pi^2}{GM} \cdot a^3 \right)^{1/2}$$

$$= \left(\frac{4\pi^2}{(6.637 \times 10^{-8})(2 \times 10^{33})} \cdot (2.87 \times 10^{14})^3 \right)^{1/2}$$

$$\approx [70.308 \times 10^{17}]^{1/2} \text{ sec}$$

$$= 26.516 \times 10^8 \text{ sec}.$$

Notice how the units mesh perfectly so that our answer is in seconds. There are 3.16×10^7 seconds in an Earth year. We divide by this number to find that the time of one orbit is

$$T \approx 83.9 \text{ Earth years.}$$

Math Note: Kepler elicited his three laws of planetary motion by studying reams of observational data that had been compiled by his teacher Tycho Brahe. It was a revelation, and a virtuoso application of the analytical arts, when Newton determined how to derive the three laws logically from the universal law of gravitation. Newton himself attached little significance to the feat. He in fact lost his notes and forgot about the whole matter. It was only years later, when his friend Edmund Halley dragged the information out of him, that Newton went back to first principles and reconstructed the arguments so that he could share them with his colleagues.

2.7 Higher-Order Linear Equations, Coupled Harmonic Oscillators

We treat here some aspects of higher-order equations that bear a similarity to what we learned about second-order examples. We shall concentrate primarily on linear equations with constant coefficients. As usual, we illustrate the ideas with a few key examples.

Math Note: One of the pleasant features of the linear theory of ordinary differential equations is that the higher-order theory very strongly resembles the second-order theory. Such is not the case for nonlinear equations. In that context there is much less coherence of the ideas.

We consider an equation of the form

$$y^{(n)} + a_{n-1}y^{(n-1)} + \cdots + a_1 y^{(1)} + a_0 y = f. \tag{1}$$

Here a superscript (j) denotes a jth derivative and f is some continuous function. This is a linear, ordinary differential equation of order n.

Following what we learned about second-order equations, we expect the general solution of (1) to have the form

$$y = y_p + y_g,$$

where y_p is a particular solution of (1) and y_g is the general solution of the associated homogeneous equation

$$y^{(n)} + a_{n-1}y^{(n-1)} + \cdots + a_1 y^{(1)} + a_0 y = 0. \tag{2}$$

Furthermore, we expect that y_g will have the form

$$y_g = A_1 y_1 + A_2 y_2 + \cdots + A_{n-1}y_{n-1} + A_n y_n,$$

where the y_j are "independent" solutions of (2).

We begin by studying the homogeneous equation (2) and seeking the general solution y_g. Again following the paradigm that we developed for second-order equations, we guess a solution of the form $y = e^{rx}$. Substituting this guess into (2), we find that

$$e^{rx} \cdot \left[r^n + a_{n-1}r^{n-1} + \cdots + a_1 r + a_0 \right] = 0.$$

Thus we are led to solving the *associated polynomial*

$$r^n + a_{n-1}r^{n-1} + \cdots + a_1 r + a_0 = 0.$$

The fundamental theorem of algebra tells us that every polynomial of degree n has a total of n complex roots $r_1, r_2, \ldots, r_n$ (there may be repetitions in this list). Thus the polynomial factors are

$$(r - r_1) \cdot (r - r_2) \cdots (r - r_{n-1}) \cdot (r - r_n).$$

In practice there may be some difficulty in *actually finding* the complete set of roots of a given polynomial. For instance, it is known that for polynomials of degree 5 and greater there is no elementary formula for the roots. Let us pass over this sticky point for the moment, and continue to comment on the theoretical setup.

I. Distinct Real Roots: For a given associated polynomial, if the roots $r_1, r_2, \ldots, r_n$ are distinct and real, then we can be sure that

$$e^{r_1 x}, e^{r_2 x}, \ldots, e^{r_n x}$$

are n distinct solutions to the differential equation (2). It then follows, just as in the second-order case, that

$$y_g = A_1 e^{r_1 x} + A_2 e^{r_2 x} + \cdots + A_n e^{r_n x}$$

is the general solution to (2) that we seek.

II. Repeated Real Roots: If the roots are real, but two of them are equal (say that $r_1 = r_2$), then of course $e^{r_1 x}$ and $e^{r_2 x}$ are *not* distinct solutions of the differential equation. Just as in the case of second-order equations, what we do in this case is manufacture two distinct solutions of the form $e^{r_1 x}$ and $x \cdot e^{r_1 x}$.

More generally, if several of the roots are equal, say $r_1 = r_2 = \cdots = r_k$, then we manufacture distinct solutions of the form $e^{r_1 x}, x \cdot e^{r_1 x}, x^2 \cdot e^{r_1 x}, \ldots, x^{k-1} \cdot e^{r_1 x}$.

III. Complex Roots: We have been assuming that the coefficients of the original differential equation ((1) or (2)) are all real. This being the case, any complex roots of the associated polynomial will occur in conjugate pairs $a + ib$ and $a - ib$. Then we have distinct solutions $e^{(a+ib)x}$ and $e^{(a-ib)x}$. Now we can use Euler's formula and a little algebra, just as we did in the second-order case, to produce distinct real solutions $e^{ax} \cos bx$ and $e^{ax} \sin bx$.

In the case that complex roots are repeated to order k, then we take

$$e^{ax} \cos bx, xe^{ax} \cos bx, \ldots, x^{k-1} e^{ax} \cos bx$$

and

$$e^{ax} \sin bx, xe^{ax} \sin bx, \ldots, x^{k-1} e^{ax} \sin bx$$

as solutions of the ordinary differential equation.

EXAMPLE 2.13

e.g.

Find the general solution of the differential equation

$$y^{(4)} - 5y^{(2)} + 4y = 0.$$

SOLUTION

The associated polynomial is

$$r^4 - 5r^2 + 4 = 0.$$

Of course we may factor this as $(r^2 - 4)(r^2 - 1) = 0$ and then as

$$(r - 2)(r + 2)(r - 1)(r + 1) = 0.$$

We find, therefore, that the general solution of our differential equation is

$$y(x) = A_1 e^{2x} + A_2 e^{-2x} + A_3 e^x + A_4 e^{-x}.$$

 EXAMPLE 2.14

Find the general solution of the differential equation

$$y^{(4)} - 8y^{(2)} + 16y = 0.$$

SOLUTION

The associated polynomial is

$$r^4 - 8r^2 + 16 = 0.$$

This factors readily as $(r^2 - 4)(r^2 - 4) = 0$, and then as

$$(r - 2)^2 (r + 2)^2 = 0.$$

According to our discussion in Part II, the general solution of the differential equation is then

$$y(x) = A_1 e^{2x} + A_2 x e^{2x} + A_3 e^{-2x} + A_4 x e^{-2x}.$$

 EXAMPLE 2.15

Find the general solution of the differential equation

$$\frac{d^4 y}{dx^4} - 2\frac{d^3 y}{dx^3} + 2\frac{d^2 y}{dx^2} - 2\frac{dy}{dx} + y = 0.$$

SOLUTION

The associated polynomial is

$$r^4 - 2r^3 + 2r^2 - 2r + 1 = 0.$$

We notice, just by inspection, that $r_1 = 1$ is a solution of this polynomial equation. Thus $r - 1$ divides the polynomial. In fact

$$r^4 - 2r^3 + 2r^2 - 2r + 1 = (r - 1) \cdot (r^3 - r^2 + r - 1).$$

But we again see that $r_2 = 1$ is a root of the new third-degree polynomial. Dividing out $r - 1$ again, we obtain a quadratic polynomial that we can solve directly.

The end result is

$$r^4 - 2r^3 + 2r^2 - 2r + 1 = (r-1)^2 \cdot (r^2 + 1) = 0$$

or

$$(r-1)^2(r-i)(r+i) = 0.$$

As a result, we find that the general solution of the differential equation is

$$y(x) = A_1 e^x + A_2 x e^x + A_3 \cos x + A_4 \sin x.$$

EXAMPLE 2.16
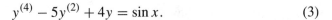

Find the general solution of the equation

$$y^{(4)} - 5y^{(2)} + 4y = \sin x. \tag{3}$$

SOLUTION
In fact we found the general solution of the associated homogeneous equation in Example 2.13. To find a particular solution of (3), we use undetermined coefficients and guess a solution of the form $y = \alpha \cos x + \beta \sin x$. A little calculation reveals then that $y_p(x) = (1/10) \sin x$ is the particular solution that we seek. As a result,

$$y(x) = \frac{1}{10} \sin x + A_1 e^{2x} + A_2 e^{-2x} + A_3 e^x + A_4 e^{-x}$$

is the general solution of (3).

You Try It: Find the general solution of the differential equation

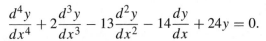

$$\frac{d^4 y}{dx^4} + 2\frac{d^3 y}{dx^3} - 13\frac{d^2 y}{dx^2} - 14\frac{dy}{dx} + 24y = 0.$$

[*Hint:* The associated polynomial is $r^4 + 2r^3 - 13r^2 - 14r + 24 = 0$. The rational roots of this polynomial will be factors of the constant term 24.]

EXAMPLE 2.17 (Coupled Harmonic Oscillators)

Linear equations of order greater than two arise in physics by the elimination of variables from simultaneous systems of second-order equations. We give here an example that arises from coupled harmonic oscillators. Accordingly, let two carts of masses m_1, m_2 be attached to left and right walls as in Fig. 2.13 with springs having spring constants k_1, k_2. If there is no damping and the the carts are unattached, then of course when the carts are perturbed we have two separate harmonic oscillators.

But if we connect the carts, with a spring having spring constant k_3, then we obtain *coupled harmonic oscillators*. In fact Newton's second law of motion

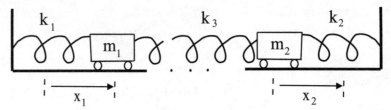

Fig. 2.13.

can now be used to show that the motions of the coupled carts will satisfy these differential equations:

$$m_1 \frac{d^2 x_1}{dt^2} = -k_1 x_1 + k_3 (x_2 - x_1),$$

$$m_2 \frac{d^2 x_2}{dt^2} = -k_2 x_2 - k_3 (x_2 - x_1).$$

We can solve the first equation for x_2,

$$x_2 = \frac{1}{k_3} \left(x_1 [k_1 + k_3] + m_1 \frac{d^2 x_1}{dt^2} \right),$$

and then substitute into the second equation. The result is a fourth-order equation for x_1.

Exercises

1. Find the general solution of each of the following differential equations:
   (a) $y'' + y' - 6y = 0$
   (b) $y'' + 2y' + y = 0$

2. Find the solution of each of the following initial value problems:
   (a) $y'' - 5y' + 6y = 0$, $y(1) = e^2$ and $y'(1) = 3e^2$
   (b) $y'' - 6y' + 5y = 0$, $y(0) = 3$ and $y'(0) = 11$

3. Find the differential equation of each of the following general solution sets:
   (a) $Ae^x + Be^{-2x}$
   (b) $A + Be^{2x}$

4. Use the method of variation of parameters to find the general solution of each of the following equations:
    (a) $y'' + 3y' - 10y = 6e^{4x}$
    (b) $y'' + 4y = 3 \sin x$

5. Find a particular solution of each of the following differential equations:
    (a) $y'' - 3y = x - x^2$
    (b) $y'' + 2y' + 5y = xe^{-x}$

6. Find the general solution of each of the following equations:
    (a) $(x^2 - 1)y'' - 2xy' + 2y = x^2 + 1$
    (b) $(x^2 + x)y'' + (2x + 1)y' - \dfrac{2x + 1}{x} \cdot y = -4x^2 - 3x$

7. The equation $xy'' + 3y' = 0$ has the obvious solution $y_1 \equiv 1$. Find y_2 and find the general solution.

8. Verify that $y_1 = x^2$ is one solution of $x^2 y'' + xy' - 4y = 0$, and then find y_2 and the general solution.

9. Find the general solution of the differential equation:
    (a) $y'' + 2y' + 4y = 0$
    (b) $y'' - 3y' + 6y = x$

10. In each problem, find the general solution of the given differential equation:
    (a) $y''' - 3y'' + 2y' = x$
    (b) $y''' - 3y'' + 4y' - 2y = 0$

11. The planet Zulu describes an elliptical orbit about the sun. It is known that the semi-major axis of this orbit has length 1200×10^6 kilometers. The gravitational constant is $G = 6.637 \times 10^{-8}$ cm^3/($g \cdot$ sec^2). Finally, the mass of the sun is 2×10^{33} grams. Determine the period of the orbit of Zulu.

CHAPTER 3

Power Series Solutions and Special Functions

3.1 Introduction and Review of Power Series

It is useful to classify the functions that we know, or will soon know, in an informal way. The *polynomials* are functions of the form

$$a_0 + a_1 x + a_2 x^2 + \cdots + a_{n-1} x^{n-1} + a_n x^n,$$

where $a_0, a_1, \ldots, a_n$ are constants. This is a polynomial of degree n.

A *transcendental function* is one that is not a polynomial. The *elementary transcendental functions* are the ones that we encounter in calculus class: sine, cosine, logarithm, exponential, and their inverses and combinations using arithmetic/algebraic operations.

EXAMPLE 3.1

Calculate the radius of convergence of the series

$$\sum_{j=0}^{\infty} \frac{x^j}{j^2}.$$

SOLUTION

We apply the ratio test:

$$\lim_{j\to\infty} \left| \frac{x^{j+1}/(j+1)^2}{x^j/j^2} \right| = \left| \lim_{j\to\infty} \frac{j^2}{(j+1)^2} \cdot x \right| = |x|.$$

We know that the series will converge when this limit is less than 1, or $|x| < 1$. Likewise, it diverges when $|x| > 1$. Thus the radius of convergence is $R = 1$.

In practice, one has to check the *endpoints* of the interval of convergence by hand for each case. In this example, we see immediately that the series converges at ± 1. Thus we may say that the interval of convergence is $[-1, 1]$.

EXAMPLE 3.2

Calculate the radius of convergence of the series

$$\sum_{j=0}^{\infty} \frac{x^j}{j}.$$

SOLUTION

We apply the ratio test:

$$\lim_{j\to\infty} \left| \frac{x^{j+1}/j+1}{x^j/j} \right| = \left| \lim_{j\to\infty} \frac{j}{j+1} \cdot x \right| = |x|.$$

We know that the series will converge when this limit is less than 1, or $|x| < 1$. Likewise, it converges when $|x| > 1$. Thus the radius of convergence is $R = 1$.

In this example, we see immediately that the series converges at -1 (by the alternating series test) and diverges at $+1$ (since this gives the harmonic series). Thus we may say that the interval of convergence is $[-1, 1)$.

EXAMPLE 3.3

Calculate the radius of convergence of the series

$$\sum_{j=0}^{\infty} \frac{x^j}{j^j}.$$

SOLUTION
We use the root test:

$$\lim_{j \to \infty} \left| \frac{x^j}{j^j} \right|^{1/j} = \lim_{j \to \infty} \left| \frac{x}{j} \right| = 0.$$

Of course $0 < 1$, regardless of the value of x. So the series converges for all x. The radius of convergence is $+\infty$ and the interval of convergence is $(-\infty, +\infty)$. There is no need to check the endpoints of the interval of convergence, because there are none.

☞ **You Try It:**   Calculate the interval of convergence for the power series

$$\sum_{j=0}^{\infty} \frac{x^{2j}}{(2j)!}.$$

III.   Suppose that our power series (1) converges for $|x| < R$ with $R > 0$. Denote its sum by $f(x)$, so

$$f(x) = \sum_{j=0}^{\infty} a_j x^j = a_0 + a_1 x + a_2 x^2 + \cdots .$$

Thus the power series defines a *function*, and we may consider differentiating it. In fact the function f is continuous and has derivatives of all orders. We may calculate the derivatives by differentiating the series termwise:

$$f'(x) = \sum_{j=1}^{\infty} j a_j x^{j-1} = a_1 + 2a_2 x + 3a_3 x^2 + \cdots ,$$

$$f''(x) = \sum_{j=2}^{\infty} j(j-1) x^{j-2} = 2a_2 + 3 \cdot 2a_3 x^2 + \cdots ,$$

and so forth. Each of these series converges on the same interval $|x| < R$.

Observe that if we evaluate the first of these formulas at $x = 0$, then we obtain the useful fact that

$$a_1 = \frac{f'(0)}{1!}.$$

If we evaluate the second formula at $x = 0$, then we obtain the analogous fact that

$$a_2 = \frac{f^{(2)}(0)}{2!}.$$

In general, we can derive (by successive differentiation) the formula

$$a_j = \frac{f^{(j)}(0)}{j!}, \tag{2}$$

which gives us an explicit way to determine the coefficients of the power series expansion of a function. It follows from these considerations that a power series is identically equal to 0 if and only if each of its coefficients is 0.

We may also note that a power series may be integrated termwise. If

$$f(x) = \sum_{j=0}^{\infty} a_j x^j = a_0 + a_1 x + a_2 x^2 + \cdots,$$

then

$$\int f(x)\,dx = \sum_{j=0}^{\infty} a_j \frac{x^{j+1}}{j+1} = a_0 x + a_1 \frac{x^2}{2} + a_2 \frac{x^3}{3} + \cdots.$$

If

$$f(x) = \sum_{j=0}^{\infty} a_j x^j = a_0 + a_1 x + a_2 x^2 + \cdots$$

and

$$g(x) = \sum_{j=0}^{\infty} b_j x^j = b_0 + b_1 x + b_2 x^2 + \cdots$$

for $|x| < R$, then these functions may be added or subtracted by adding or subtracting the series termwise:

$$f(x) \pm g(x) = \sum_{j=0}^{\infty} (a_j \pm b_j) x^j = (a_0 \pm b_0) + (a_1 \pm b_1)x + (a_2 \pm b_2)x^2 + \cdots.$$

Also f and g may be multiplied as if they were polynomials, to wit

$$f(x) \cdot g(x) = \sum_{j=0}^{\infty} c_n x^n,$$

where

$$c_n = a_0 b_n + a_1 b_{n-1} + \cdots + a_n b_0.$$

We shall say more about operations on power series below.

Finally, we note that if two different power series converge to the same function, then (2) tells us that the two series are precisely the same (i.e., have the same coefficients). In particular, if $f(x) \equiv 0$ for $|x| < R$, then all the coefficients of the power series expansion for f are equal to 0.

IV. Suppose that f is a function that has derivatives of all orders on $|x| < R$. We may calculate the coefficients

$$a_j = \frac{f^{(j)}(0)}{j!}$$

and then write the (formal) series

$$\sum_{j=0}^{\infty} a_j x^j. \tag{3}$$

It is then natural to ask whether the series (3) *converges to* f. When the function f is sine or cosine or logarithm or the exponential, then the answer is "yes." But these are very special functions. Actually, the answer to our question is generically "no." Most infinitely differentiable functions do *not* have power series expansion that converges back to the function. In fact most have power series that does not converge at all; but even in the unlikely circumstance that the series does converge, it will most probably *not* converge to the original f.

This circumstance may seem rather strange, but it explains why mathematicians spent so many years trying to understand power series. The functions that *do* have convergent power series are called *real analytic* and they are very particular functions with remarkable properties. Even though the subject of real analytic functions is more than 300 years old, the first and only book written on the subject is [KRP1].

We do have a way of coming to grips with the unfortunate state of affairs that has just been described, and that is the theory of *Taylor expansions*. For a function with sufficiently many derivatives, here is what is actually true:

$$f(x) = \sum_{j=0}^{n} \frac{f^{(j)}(0)}{j!} x^j + R_n(x), \tag{4}$$

where the remainder term $R_n(x)$ is given by

$$R_n(x) = \frac{f^{(n+1)}(\xi)}{(n+1)!} x^{n+1}$$

for some number ξ between 0 and x. The power series converges to f precisely when the partial sums in (4) converge, and that happens precisely when the remainder term goes to zero. What is important for you to understand is that, generically, the remainder term *does not go to zero*. But formula (4) is still valid.

We can use formula (4) to obtain the familiar power series expansions for several important functions:

$$e^x = \sum_{j=0}^{\infty} \frac{x^j}{j!} = 1 + x + \frac{x^2}{2!} + \frac{x^3}{3!} + \cdots ;$$

$$\sin x = \sum_{j=0}^{\infty} (-1)^j \frac{x^{2j+1}}{(2j+1)!} = x - \frac{x^3}{3!} + \frac{x^5}{5!} - + \cdots ;$$

$$\cos x = \sum_{j=0}^{\infty} (-1)^j \frac{x^{2j}}{(2j)!} = 1 - \frac{x^2}{2!} + \frac{x^4}{4!} - + \cdots .$$

Of course there are many others, including the logarithm and the other trigonometric functions. Just for practice, let us verify that the first of these formulas is actually valid.

First,

$$\frac{d^j}{dx^j} e^x = e^x \qquad \text{for every } j.$$

Thus

$$a_j = \frac{[d^j/dx^j]e^x\big|_{x=0}}{j!} = \frac{1}{j!}.$$

This confirms that the formal power series for e^x is just what we assert it to be. To check that it converges back to e^x, we must look at the remainder term, which is

$$R_n(x) = \frac{f^{(n+1)}(\xi)}{(n+1)!} x^{n+1} = \frac{e^\xi \cdot x^{n+1}}{(n+1)!}.$$

Of course, for x fixed, we have that $|\xi| < |x|$; also $n \to \infty$ implies that $(n+1)! \to \infty$ much faster than $x^{n+1} \to \infty$. So the remainder term goes to zero and the series converges to e^x.

Math Note: There are many different techniques for expanding a function into a series of basic elements. Certainly power series is one of the most important of these. In Chapter 4 we shall learn about Fourier series, which is another important methodology. One of several reasons that Fourier series are attractive is that the Fourier series of a function usually converges, and usually converges back to the original function.

V. Operations on Series Some operations on series, such as addition, subtraction, and scalar multiplication, are straightforward. Others, such as multiplication, entail subtleties.

Sums and Scalar Products of Series

PROPOSITION 3.1

Let

$$\sum_{j=1}^{\infty} a_j \quad and \quad \sum_{j=1}^{\infty} b_j$$

be convergent series of real or complex numbers; assume that the series sum to limits α and β respectively. Then

(a) *The series $\sum_{j=1}^{\infty}(a_j + b_j)$ converges to the limit $\alpha + \beta$.*
(b) *If c is a constant, then the series $\sum_{j=1}^{\infty} c \cdot a_j$ converges to $c \cdot \alpha$.*

Products of Series

In order to keep our discussion of multiplication of series as straightforward as possible, we deal at first with absolutely convergent series. It is convenient in this discussion to begin our sum at $j = 0$ instead of $j = 1$. If we wish to multiply

$$\sum_{j=0}^{\infty} a_j \quad and \quad \sum_{j=0}^{\infty} b_j,$$

then we need to specify what the partial sums of the product series should be. An obvious necessary condition that we wish to impose is that if the first series converges to α and the second converges to β, then the product series $\sum_{j=0}^{\infty} c_j$, whatever we define it to be, should converge to $\alpha \cdot \beta$.

The Cauchy Product

Cauchy's idea was that the terms for the product series should be

$$c_m \equiv \sum_{j=0}^{m} a_j \cdot b_{m-j}.$$

This particular form for the summands can be easily motivated using power series considerations (which we shall provide later on). For now we concentrate on confirming that this "Cauchy product" of two series really works.

THEOREM 3.1   *Given*

Let $\sum_{j=0}^{\infty} a_j$ and $\sum_{j=0}^{\infty} b_j$ be two absolutely convergent series which converge to limits α and β respectively. Define the series $\sum_{m=0}^{\infty} c_m$ with summands

$$c_m = \sum_{j=0}^{m} a_j \cdot b_{m-j}.$$

Then the series $\sum_{m=0}^{\infty} c_m$ converges to $\alpha \cdot \beta$.

EXAMPLE 3.4   *e.g.*

Consider the Cauchy product of the two conditionally convergent series

$$\sum_{j=0}^{\infty} \frac{(-1)^j}{\sqrt{j+1}} \quad \text{and} \quad \sum_{j=0}^{\infty} \frac{(-1)^j}{\sqrt{j+1}}.$$

Observe that

$$c_m = \frac{(-1)^0(-1)^m}{\sqrt{1}\sqrt{m+1}} + \frac{(-1)^1(-1)^{m-1}}{\sqrt{2}\sqrt{m}} + \cdots$$

$$+ \frac{(-1)^m(-1)^0}{\sqrt{m+1}\sqrt{1}}$$

$$= \sum_{j=0}^{m} (-1)^m \frac{1}{\sqrt{(j+1)\cdot(m+1-j)}}.$$

However,

$$(j+1)\cdot(m+1-j) \le (m+1)\cdot(m+1) = (m+1)^2.$$

Thus

$$|c_m| \ge \sum_{j=0}^{m} \frac{1}{m+1} = 1.$$

We thus see that the terms of the series $\sum_{m=0}^{\infty} c_m$ do not tend to zero, so the series cannot converge.

EXAMPLE 3.5   *e.g.*

The series

$$A = \sum_{j=0}^{\infty} 2^{-j} \quad \text{and} \quad B = \sum_{j=0}^{\infty} 3^{-j}$$

are both absolutely convergent. We challenge you to calculate the Cauchy product and to verify that that product converges to 3.

VI.   We conclude by summarizing some properties of real analytic functions:

1.   Polynomials and the functions e^x, $\sin x$, $\cos x$ are all real analytic at all points.
2.   If f and g are real analytic at x_0, then $f \pm g$, $f \cdot g$, and f/g (provided $g(x_0) \neq 0$) are real analytic at x_0.
3.   If f is real analytic at x_0 and if f^{-1} is a continuous inverse and $f'(x_0) \neq 0$, then f^{-1} is real analytic at $f(x_0)$.
4.   If g is real analytic at x_0 and f is real analytic at $g(x_0)$, then $f \circ g$ is real analytic at x_0.
5.   The function defined by the sum of a power series is real analytic at all interior points of its interval of convergence.

3.2 Series Solutions of First-Order Differential Equations

Now we get our feet wet and use power series to solve first-order linear equations. This will turn out to be misleadingly straightforward to do, but it will show us the basic moves.

   EXAMPLE 3.6

Solve the differential equation

$$y' = y$$

using the method of power series.

SOLUTION

Of course we already know that the solution to this equation is $y = C \cdot e^x$, but let us pretend that we do not know this. We proceed by *guessing* that the equation has a solution given by a power series, and we proceed to solve for the coefficients of that power series.

So our guess is a solution of the form

$$y = a_0 + a_1 x + a_2 x^2 + a_3 x^3 + \cdots .$$

Then

$$y' = a_1 + 2a_2 x + 3a_3 x^2 + \cdots$$

and we may substitute these two expressions into the differential equation. Thus

$$a_1 + 2a_2x + 3a_3x^2 + \cdots = a_0 + a_1x + a_2x^2 + \cdots .$$

Now the powers of x must match up (i.e., the coefficients must be equal). We conclude that

$$a_1 = a_0$$

$$2a_2 = a_1$$

$$3a_3 = a_2$$

and so forth. Let us take a_0 to be an unknown constant C. Then we see that

$$a_1 = C;$$

$$a_2 = \frac{C}{2};$$

$$a_3 = \frac{C}{3 \cdot 2}; \text{ etc.}$$

In general,

$$a_n = \frac{C}{n!}.$$

In summary, our power series solution of the original differential equation is

$$y = \sum_{j=0}^{\infty} \frac{C}{j!}x^j = C \cdot \sum_{j=0}^{\infty} \frac{x^j}{j!} = C \cdot e^x.$$

Thus we have a new way, using power series, of discovering the general solution of the differential equation $y' = y$.

The next example illustrates the point that, by running our logic a bit differently, we can *use* a differential equation to derive the power series expansion for a given function.

EXAMPLE 3.7

Let p be an arbitrary real constant. Use a differential equation to derive the power series expansion for the function

$$y = (1 + x)^p.$$

SOLUTION

Of course the given y is a solution of the initial value problem

$$(1 + x) \cdot y' = py, \quad y(0) = 1.$$

We assume that the equation has a power series solution

$$y = \sum_{j=0}^{\infty} a_j x^j = a_0 + a_1 x + a_2 x^2 + \cdots$$

with positive radius of convergence R. Then

$$y' = \sum_{j=1}^{\infty} j \cdot a_j x^{j-1} = a_1 + 2a_2 x + 3a_3 x^2 + \cdots \, ;$$

$$xy' = \sum_{j=1}^{\infty} j \cdot a_j x^j = a_1 x + 2a_2 x^2 + 3a_3 x^3 + \cdots \, ;$$

$$py = \sum_{j=0}^{\infty} p a_j x^j = p a_0 + p a_1 x + p a_2 x^2 + \cdots .$$

By the differential equation, the sum of the first two of these series equals the third. Thus

$$\sum_{j=1}^{\infty} j a_j x^{j-1} + \sum_{j=1}^{\infty} j a_j x^j = \sum_{j=0}^{\infty} p a_j x^j.$$

We immediately see two interesting anomalies: the powers of x on the left-hand side do not match up, so the two series cannot be immediately added. Also the summations do not all begin in the same place. We address these two concerns as follows.

First, we can change the index of summation in the first sum on the left to obtain

$$\sum_{j=0}^{\infty} (j + 1) a_{j+1} x^j + \sum_{j=1}^{\infty} j a_j x^j = \sum_{j=0}^{\infty} p a_j x^j.$$

Write out the first few terms of the sum we have changed, and the original sum, to see that they are just the same.

Now every one of our series has x^j in it, but they begin at different places. So we break off the extra terms as follows:

$$\sum_{j=1}^{\infty}(j+1)a_{j+1}x^j + \sum_{j=1}^{\infty} ja_j x^j - \sum_{j=1}^{\infty} pa_j x^j = -a_1 x^0 - pa_0 x^0. \qquad (1)$$

Notice that all we have done is to break off the zeroth terms of the first and third series, and put them on the right.

The three series on the left-hand side of (1) are begging to be put together: they have the same form, they all involve powers of x, and they all begin at the same index. Let us do so:

$$\sum_{j=1}^{\infty}[(j+1)a_{j+1} + ja_j - pa_j]x^j = -a_1 + pa_0.$$

Now the powers of x that appear on the left are 1, 2, ..., and there are none of these on the right. We conclude that each of the coefficients on the left is zero; by the same reasoning, the coefficient $(-a_1 + pa_0)$ on the right (i.e., the constant term) equals zero. So we have the equations[2]

$$-a_1 + pa_0 = 0$$

$$(j+1)a_{j+1} + (j-p)a_j = 0 \quad \text{for } j = 1, 2, \ldots.$$

Our initial condition tells us that $a_0 = 1$. Then our first equation implies that $a_1 = p$. The next equation, with $j = 1$, says that

$$2a_2 + (1-p)a_1 = 0.$$

Hence $a_2 = (p-1)a_1/2 = (p-1)p/2$. Continuing, we take $j = 2$ in the second equation to get

$$3a_3 + (2-p)a_2 = 0$$

so $a_3 = (p-2)a_2/3 = (p-2)(p-1)p/(3 \cdot 2)$.

We may continue in this manner to obtain that

$$a_j = \frac{p(p-1)(p-2)\cdots(p-j+1)}{j!}.$$

---

[2]A set of equations like this is called a *recursion*. It expresses later indexed a_j's in terms of earlier indexed a_j's.

Thus the power series expansion for our solution y is

$$y = 1 + px + \frac{p(p-1)}{2!}x + \frac{p(p-1)(p-2)}{3!} + \cdots$$
$$+ \frac{p(p-1)(p-2)\cdots(p-j+1)}{j!}x^j + \cdots.$$

Since we knew in advance that the solution of our initial value problem was

$$y = (1+x)^p,$$

we find that we have derived Isaac Newton's general binomial theorem (or binomial series):

$$(1+x)^p = 1 + px + \frac{p(p-1)}{2!}x + \frac{p(p-1)(p-2)}{3!} + \cdots$$
$$+ \frac{p(p-1)(p-2)\cdots(p-j+1)}{j!}x^j + \cdots.$$

You Try It: Use power series methods to solve the differential equation

$$y' = xy.$$

 Math Note: It is tempting to use the power series method to attack virtually any differential equation that we encounter. But the method only works when the coefficients of the differential equation are themselves real analytic. And it works best for linear equations.

3.3 Second-Order Linear Equations: Ordinary Points

We have invested considerable effort in studying equations of the form

$$y'' + p \cdot y' + q \cdot y = 0. \tag{1}$$

In some sense, our investigations thus far have been misleading; for we have only considered particular equations in which a closed-form solution can be found. These cases are really the exception rather than the rule. For most such equations, there is no "formula" for the solution. Power series then give us some extra flexibility. Now we may seek a power series solution; that solution is valid, and may be calculated and used in applications, even though it may not be expressed in a compact formula.

A number of the differential equations that arise in mathematical physics—Bessel's equation, Lagrange's equation, and many others—in fact fit the description that we have just presented. So it is worthwhile to develop techniques for studying (1). In the present section we shall concentrate on finding a power series solution to the equation (1)—written in standard form—expanded about a point x_0, where x_0 has the property that p and q have convergent power series expansions about x_0. In this circumstance we call x_0 an *ordinary point* of the differential equation.

We begin our study with a familiar equation, just to see the basic steps, and how the solution will play out.

EXAMPLE 3.8

Solve the differential equation

$$y'' + y = 0$$

by power series methods.

SOLUTION

As usual, we guess a solution of the form

$$y = \sum_{j=0}^{\infty} a_j x^j = a_0 + a_1 x + a_2 x^2 + \cdots .$$

Of course it follows that

$$y' = \sum_{j=1}^{\infty} j a_j x^{j-1} = a_1 + 2a_2 x + 3a_3 x^2 + \cdots$$

and

$$y'' = \sum_{j=2}^{\infty} j(j-1) a_j x^{j-2} = 2 \cdot 1 \cdot a_2 + 3 \cdot 2 \cdot 1 \cdot a_3 x + 4 \cdot 3 \cdot 2 \cdot 1 \cdot x^2 \cdots .$$

Plugging the first and third of these into the differential equation gives

$$\sum_{j=2}^{\infty} j(j-1) a_j x^{j-2} + \sum_{j=0}^{\infty} a_j x^j = 0.$$

As in the last example of Section 3.2, we find that the series have x raised to different powers, and that the summation begins in different places. We follow the standard procedure for repairing these matters.

First, we change the index of summation in the second series. So

$$\sum_{j=2}^{\infty} j(j-1)a_j x^{j-2} + \sum_{j=2}^{\infty} a_{j-2} x^{j-2} = 0.$$

We invite you to verify that the new second series is just the same as the original second series (merely write out a few terms of each to check). We are fortunate in that both series now begin at the same index. So we may add them together to obtain

$$\sum_{j=2}^{\infty} [j(j-1)a_j + a_{j-2}] x^{j-2} = 0.$$

The only way that such a power series can be identically zero is if each of the coefficients is zero. So we obtain the recursion equations

$$j(j-1)a_j + a_{j-2} = 0, \qquad j = 2, 3, 4, \ldots.$$

Then $j = 2$ gives us

$$a_2 = -\frac{a_0}{2 \cdot 1}.$$

It will be convenient to take a_0 to be an arbitrary constant A, so that

$$a_2 = -\frac{A}{2 \cdot 1}.$$

The recursion for $j = 4$ says that

$$a_4 = -\frac{a_2}{4 \cdot 3} = \frac{A}{4 \cdot 3 \cdot 2 \cdot 1}.$$

Continuing in this manner, we find that

$$a_{2j} = (-1)^j \cdot \frac{A}{2j \cdot (2j-1) \cdot (2j-2) \cdots 3 \cdot 2 \cdot 1}$$

$$= (-1)^j \cdot \frac{A}{(2j)!}, \qquad j = 1, 2, \ldots.$$

Thus we have complete information about the coefficients with even index.

Now let us consider the odd indices. Look at the recursion for $j = 3$. This is

$$a_3 = -\frac{a_1}{3 \cdot 2}.$$

It is convenient to take a_1 to be an arbitrary constant B. Thus

$$a_3 = -\frac{B}{3 \cdot 2 \cdot 1}.$$

Continuing with $j = 5$, we find that

$$a_5 = -\frac{a_3}{5 \cdot 4} = \frac{B}{5 \cdot 4 \cdot 3 \cdot 2 \cdot 1}.$$

In general,

$$a_{2j+1} = (-1)^j \frac{B}{(2j+1)!}, \qquad j = 1, 2, \ldots.$$

In summary then, the general solution of our differential equation is given by

$$y = A \cdot \left(\sum_{j=0}^{\infty} (-1)^j \cdot \frac{A}{(2j)!} x^{2j} \right) + B \cdot \left(\sum_{j=0}^{\infty} (-1)^j \frac{B}{(2j+1)!} x^{2j+1} \right).$$

Of course we recognize the first power series as the cosine function and the second as the sine function. So we have rediscovered that the general solution of $y'' + y = 0$ is

$$y = A \cdot \cos x + B \cdot \sin x.$$

EXAMPLE 3.9

e.g.

Use the method of power series to solve the differential equation

$$(1 - x^2)y'' - 2xy' + p(p+1)y = 0. \qquad (2)$$

Here p is an arbitrary real constant. This is called *Legendre's equation*.

SOLUTION

First we write the equation in standard form:

$$y'' - \frac{2x}{1-x^2}y' + \frac{p(p+1)}{1-x^2} = 0.$$

Observe that, near $x = 0$, division by 0 is avoided and the coefficients p and q are real analytic. So 0 is an ordinary point.

We therefore guess a solution of the form

$$y = \sum_{j=0}^{\infty} a_j x^j = a_0 + a_1 x + a_2 x^2 + \cdots$$

and calculate

$$y' = \sum_{j=1}^{\infty} j a_j x^{j-1} = a_1 + 2a_2 x + 3a_3 x^2 + \cdots$$

and

$$y'' = \sum_{j=2}^{\infty} j(j-1) a_j x^{j-2} = 2a_2 + 3 \cdot 2 \cdot a_3 x + \cdots.$$

It is most convenient to treat the differential equation in the form (2). We calculate

$$-x^2 y'' = -\sum_{j=2}^{\infty} j(j-1) a_j x^j$$

and

$$-2xy' = -\sum_{j=1}^{\infty} 2 j a_j x^j.$$

Substituting into the differential equation now yields

$$\sum_{j=2}^{\infty} j(j-1) a_j x^{j-2} - \sum_{j=2}^{\infty} j(j-1) a_j x^j$$

$$- \sum_{j=1}^{\infty} 2 j a_j x^j + p(p+1) \sum_{j=0}^{\infty} a_j x^j = 0.$$

We adjust the index of summation in the first sum so that it contains x^j rather than x^{j-2} and we break off spare terms and collect them on the right. The result is

$$\sum_{j=2}^{\infty} (j+2)(j+1) a_{j+2} x^j - \sum_{j=2}^{\infty} j(j-1) a_j x^j$$

$$- \sum_{j=2}^{\infty} 2 j a_j x^j + p(p+1) \sum_{j=2}^{\infty} a_j x^j$$

$$= -2a_2 - 6a_3 x + 2a_1 x - p(p+1) a_0 - p(p+1) a_1 x.$$

In other words,

$$\sum_{j=2}^{\infty} \left[(j+2)(j+1)a_{j+2} - j(j-1)a_j - 2ja_j + p(p+1)a_j \right] x^j$$

$$= -2a_2 - 6a_3 x + 2a_1 x - p(p+1)a_0 - p(p+1)a_1 x.$$

As a result,

$$\left[(j+2)(j+1)a_{j+2} - j(j-1)a_j - 2ja_j + p(p+1)a_j \right] = 0,$$

$$\text{for } j = 2, 3, \ldots$$

together with

$$-2a_2 - p(p+1)a_0 = 0$$

and

$$-6a_3 + 2a_1 - p(p+1)a_1 = 0.$$

We have arrived at the recursion

$$a_2 = -\frac{p(p+1)}{1 \cdot 2} \cdot a_0,$$

$$a_3 = -\frac{(p-1)(p+2)}{2 \cdot 3} \cdot a_1,$$

$$a_{j+2} = -\frac{(p-j)(p+j+1)}{(j+2)(j+1)} \cdot a_j \qquad \text{for } j = 2, 3, \ldots. \qquad (3)$$

We recognize a familiar pattern: The coefficients a_0 and a_1 are unspecified, so we set $a_0 = A$ and $a_1 = B$. Then we may proceed to solve for the rest of the coefficients. Now

$$a_2 = -\frac{p(p+1)}{2} \cdot A,$$

$$a_3 = -\frac{(p-1)(p+2)}{2 \cdot 3} \cdot B,$$

$$a_4 = -\frac{(p-2)(p+3)}{3 \cdot 4} a_2 = \frac{p(p-2)(p+1)(p+3)}{4!} \cdot A,$$

$$a_5 = -\frac{(p-3)(p+4)}{4 \cdot 5} a_3 = \frac{(p-1)(p-3)(p+2)(p+4)}{5!} \cdot B,$$

$$a_6 = -\frac{(p-4)(p+5)}{5\cdot 6}a_4$$

$$= -\frac{p(p-2)(p-4)(p+1)(p+3)(p+5)}{6!}\cdot A,$$

$$a_7 = -\frac{(p-5)(p+6)}{6\cdot 7}a_5$$

$$= -\frac{(p-1)(p-3)(p-5)(p+2)(p+4)(p+6)}{7!}\cdot B,$$

and so forth. Putting these coefficient values into our supposed power series solution we find that the general solution of our differential equation is

$$y = A\left[1 - \frac{p(p+1)}{2!}x^2 + \frac{p(p-2)(p+1)(p+3)}{4!}x^4\right.$$

$$\left. - \frac{p(p-2)(p-4)(p+1)(p+3)(p+5)}{6!}x^6 + -\cdots\right]$$

$$+ B\left[x - \frac{(p-1)(p+2)}{3!}x^3 + \frac{(p-1)(p-3)(p+2)(p+4)}{5!}x^5\right.$$

$$\left. - \frac{(p-1)(p-3)(p-5)(p+2)(p+4)(p+6)}{7!}x^7 + -\cdots\right].$$

We assure you that, when p is not an integer, then these are *not* familiar elementary transcendental functions. These are what we call *Legendre functions*. In the special circumstance that p is a positive even integer, the first function (that which is multiplied by A) terminates as a polynomial. In the special circumstance that p is a positive odd integer, the second function (that which is multiplied by B) terminates as a polynomial. These are called *Legendre polynomials*, and they play an important role in mathematical physics, representation theory, and interpolation theory.

 You Try It: Use power series methods to solve the differential equation

$$y'' + xy = 0.$$

 Math Note: It is actually possible, without much effort, to check the radius of convergence of the functions we discovered as solutions in the last example. In fact we use the recursion relation (3) to see that

$$\left|\frac{a_{2j+2}x^{2j+2}}{a_{2j}x^{2j}}\right| = \left|-\frac{(p-2j)(p+2j+1)}{(2j+1)(2j+2)}\right|\cdot |x|^2 \to |x|^2$$

as $j \to \infty$. Thus the series expansion of the first Legendre function converges when $|x| < 1$, so the radius of convergence is 1. A similar calculation shows that the radius of convergence for the second Legendre function is 1.

We now enunciate a general result about the power series solution of an ordinary differential equation at an ordinary point.

THEOREM 3.2

Let x_0 be an ordinary point of the differential equation

$$y'' + p \cdot y' + q \cdot y = 0, \tag{4}$$

and let α and β be arbitrary real constants. Then there exists a unique real analytic function $y = y(x)$ that has a power series expansion about x_0 and so that

(a) *The function y solves the differential equation (4).*
(b) *The function y satisfies the initial conditions $y(x_0) = \alpha$, $y'(x_0) = \beta$.*

If the functions p and q have power series expansions about x_0 with radius of convergence R, then so does y.

We conclude the section with this remark. The examples that we have worked in detail resulted in solutions with *two-term* (or *binary*) recursion formulas: a_2 was expressed in terms of a_0 and a_3 was expressed in terms of a_1, etc.. In general, the recursion formulas that arise in solving an ordinary differential equation at an ordinary point may result in more complicated recursions.

Exercises

1. Use the ratio test (for example) to calculate the radius of convergence for each series:

    (a) $\sum_{j=0}^{\infty} \dfrac{2^j}{j!} x^j$

    (b) $\sum_{j=0}^{\infty} \dfrac{2^j}{3^j} x^j$

2. Verify that $R = +\infty$ for the power series expansions of sine and cosine.

3. Use Taylor's formula to confirm the validity of the power series expansions for e^x, $\sin x$, and $\cos x$.

4. (a) Show that the series

$$y = 1 - \frac{x^2}{2!} + \frac{x^4}{4!} - \frac{x^6}{6!} + - \cdots$$

satisfies $y'' = -y$.

(b) Show that the series

$$y = 1 - \frac{x^2}{2^2} + \frac{x^4}{2^2 \cdot 4^2} - \frac{x^6}{2^2 \cdot 4^2 \cdot 6^2} + - \cdots$$

converges for all x. Verify that it defines a solution of equation

$$xy'' + y' + xy = 0.$$

This function is the *Bessel function of order 0*.

5. For each of the following differential equations, find a power series solution of the form $\sum_j a_j x^j$. Endeavor to recognize this solution as the series expansion of a familiar function.

(a) $y' = 2xy$

(b) $y' + y = 1$

(c) $y' - y = 2$

6. For each of the following differential equations, find a power series solution of the form $\sum_j a_j x^j$:

(a) $xy' = y$

(b) $y' - (1/x)y = x^2$

7. Consider the equation $y'' + xy' + y = 0$.

(a) Find its general solution $y = \sum_j a_j x^j$ in the form $y = c_1 y_1(x) + c_2 y_2(x)$, where y_1, y_2 are power series.

(b) Use the ratio test to check that the two series y_1 and y_2 from part (a) converge for all x.

8. Use the method of power series to find a solution of each of these differential equations:

(a) $y'' + y = x^2$

(b) $y'' + y' = -x$

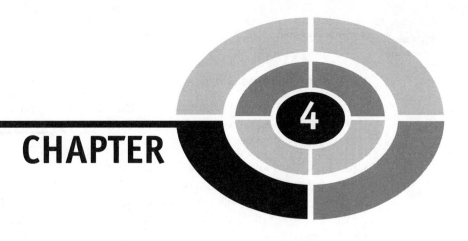

CHAPTER 4

Fourier Series: Basic Concepts

4.1 Fourier Coefficients

Trigonometric and Fourier series constitute one of the oldest parts of analysis. They arose, for instance, in classical studies of the heat and wave equations. Today they play a central role in the study of sound, heat conduction, electromagnetic waves, mechanical vibrations, signal processing, and image analysis and compression. Whereas power series (see Chapter 3) can only be used to represent very special functions (most functions, even smooth ones, do *not* have convergent power series), Fourier series can be used to represent very broad classes of functions.

For us, a trigonometric series is one of the form

$$f(x) = \frac{1}{2}a_0 + \sum_{n=1}^{\infty} \left(a_n \cos nx + b_n \sin nx \right). \tag{1}$$

We shall be concerned with two main questions:

1. Given a function f, how do we calculate the coefficients a_n, b_n?
2. Once the series for f has been calculated, can we determine that it converges, and that it converges to f?

Ultimately, we shall want to use Fourier and trigonometric series to solve ordinary and partial differential equations.

We begin our study with some classical calculations that were first performed by L. Euler (1707–1783). It is convenient to assume that our function f is defined on the interval $[-\pi, \pi] = \{x \in \mathbb{R}: -\pi \le x \le \pi\}$. We shall temporarily make the important assumption that the *trigonometric series* (1) *for f converges uniformly*. While this turns out to be true for a large class of functions (continuously differentiable functions, for example), for now this is merely a convenience so that our calculations are justified.

We apply the integral to both sides of (1). The result is

$$\int_{-\pi}^{\pi} f(x)\, dx = \int_{-\pi}^{\pi} \left[\frac{1}{2}a_0 + \sum_{n=1}^{\infty} (a_n \cos nx + b_n \sin nx) \right] dx$$

$$= \frac{1}{2} \int_{-\pi}^{\pi} a_0\, dx + \sum_{n=1}^{\infty} \int_{-\pi}^{\pi} a_n \cos nx\, dx$$

$$+ \sum_{n=1}^{\infty} \int_{-\pi}^{\pi} b_n \sin nx\, dx.$$

The change in order of summation and integration is justified by the uniform convergence of the series (see [KRA2, pp. 202 ff.]).

Now each of $\cos nx$ and $\sin nx$ integrates to 0. The result is that

$$a_0 = \frac{1}{\pi} \int_{-\pi}^{\pi} f(x)\, dx.$$

In effect, then, a_0 is the *average* of f over the interval $[-\pi, \pi]$.

To calculate a_j, we multiply the formula (1) by $\cos jx$ and then integrate as before. The result is

$$\int_{-\pi}^{\pi} f(x) \cos jx\, dx$$

$$= \int_{-\pi}^{\pi} \left[\frac{1}{2}a_0 + \sum_{n=1}^{\infty} (a_n \cos nx + b_n \sin nx) \right] \cos jx\, dx$$

$$= \int_{-\pi}^{\pi} \frac{1}{2} a_0 \cos jx \, dx$$

$$+ \sum_{n=1}^{\infty} \int_{-\pi}^{\pi} a_n \cos nx \cos jx \, dx$$

$$+ \sum_{n=1}^{\infty} \int_{-\pi}^{\pi} b_n \sin nx \cos jx \, dx. \tag{2}$$

Now the first integral on the right vanishes, as we have already noted. Further recall that

$$\cos nx \cos jx = \frac{1}{2} [\cos(n+j)x + \cos(n-j)x]$$

and

$$\sin nx \cos jx = \frac{1}{2} [\sin(n+j)x + \sin(n-j)x].$$

It follows immediately that

$$\int_{-\pi}^{\pi} \cos nx \cos jx \, dx = 0 \qquad \text{when } n \neq j$$

and

$$\int_{-\pi}^{\pi} \sin nx \cos jx \, dx = 0 \qquad \text{for all } n, j.$$

Thus our formula (2) reduces to

$$\int_{-\pi}^{\pi} f(x) \cos jx \, dx = \int_{-\pi}^{\pi} a_j \cos jx \cos jx \, dx.$$

We may use our formula above for the product of cosines to integrate the right-hand side. The result is

$$\int_{-\pi}^{\pi} f(x) \cos jx \, dx = a_j \cdot \pi$$

or

$$a_j = \frac{1}{\pi} \int_{-\pi}^{\pi} f(x) \cos jx \, dx.$$

A similar calculation shows that

$$b_j = \frac{1}{\pi} \int_{-\pi}^{\pi} f(x) \sin jx \, dx.$$

In summary, we now have formulas for calculating all the a_j's and b_j's. We display them now for reference:

$$a_0 = \frac{1}{\pi} \int_{-\pi}^{\pi} f(x)\, dx;$$

$$a_j = \frac{1}{\pi} \int_{-\pi}^{\pi} f(x) \cos jx\, dx \text{ for } j \geq 1;$$

$$b_j = \frac{1}{\pi} \int_{-\pi}^{\pi} f(x) \sin jx\, dx \text{ for } j \geq 1.$$

e.g. **EXAMPLE 4.1**

Find the Fourier series of the function

$$f(x) = x, \qquad -\pi \leq x \leq \pi.$$

SOLUTION

Of course

$$a_0 = \frac{1}{\pi} \int_{-\pi}^{\pi} x\, dx = \frac{1}{\pi} \cdot \frac{x^2}{2}\bigg|_{-\pi}^{\pi} = 0.$$

For $j \geq 1$, we calculate a_j as follows:

$$a_j = \frac{1}{\pi} \int_{-\pi}^{\pi} x \cos jx\, dx$$

$$\overset{\text{(parts)}}{=} \frac{1}{\pi} \left[x \frac{\sin jx}{j}\bigg|_{-\pi}^{\pi} - \int_{-\pi}^{\pi} \frac{\sin jx}{j}\, dx \right]$$

$$= \frac{1}{\pi} \left[0 - \left(-\frac{\cos jx}{j^2}\bigg|_{-\pi}^{\pi} \right) \right]$$

$$= 0.$$

Similarly, we calculate the b_j:

$$b_j = \frac{1}{\pi} \int_{-\pi}^{\pi} x \sin jx\, dx$$

$$\overset{\text{(parts)}}{=} \frac{1}{\pi} \left[x \frac{-\cos jx}{j}\bigg|_{-\pi}^{\pi} - \int_{-\pi}^{\pi} \frac{-\cos jx}{j}\, dx \right]$$

$$= \frac{1}{\pi} \left[-\frac{2\pi \cos j\pi}{j} - \left(-\frac{\sin jx}{j^2}\bigg|_{-\pi}^{\pi} \right) \right]$$

$$= \frac{2 \cdot (-1)^{j+1}}{j}.$$

Now that all the coefficients have been calculated, we may summarize the result as

$$x = f(x) = 2\left(\sin x - \frac{\sin 2x}{2} + \frac{\sin 3x}{3} - + \cdots\right).$$

You Try It: Calculate the Fourier series of the function

$$f(x) = 2x + 1.$$

Now calculate the Fourier series of $g(x) = \sin 2x$. Why are your answers so different?

It is convenient, in the study of Fourier series, to think of our functions as defined on the entire real line. We extend a function that is initially given on the interval $[-\pi, \pi]$ to the entire line using the idea of *periodicity*. The sine function and cosine function are periodic in the sense that $\sin(x + 2\pi) = \sin x$ and $\cos(x + 2\pi) = \cos x$. We say that sine and cosine are *periodic with period* 2π. Thus it is natural, if we are given a function f on $[-\pi, \pi)$, to define $f(x + 2\pi) = f(x)$, $f(x + 2 \cdot 2\pi) = f(x)$, $f(x - 2\pi) = f(x)$, etc.[1]

Figure 4.1 exhibits the periodic extension of the function $f(x) = x$ on $[-\pi, \pi)$ to the real line.

Figure 4.2 shows the first four summands of the Fourier series for $f(x) = x$. The finest dashes show the curve $y = 2 \sin x$, the next finest is $-\sin 2x$, the next is $(2/3) \sin 3x$, and the coarsest is $-(1/2) \sin 4x$.

Figure 4.3 shows the sum of the first four terms of the Fourier series and also of the first six terms, as compared to $f(x) = x$. Figure 4.4 shows the sum of the first eight terms of the Fourier series and also of the first ten terms, as compared to $f(x) = x$.

EXAMPLE 4.2

Calculate the Fourier series of the function

$$f(x) = \begin{cases} 0 & \text{if } -\pi \leq x < 0 \\ \pi & \text{if } 0 \leq x \leq \pi. \end{cases}$$

---

[1]Notice that we take the original function f to be defined on $[0, 2\pi)$ rather than $[0, 2\pi]$ to avoid any ambiguity at the endpoints.

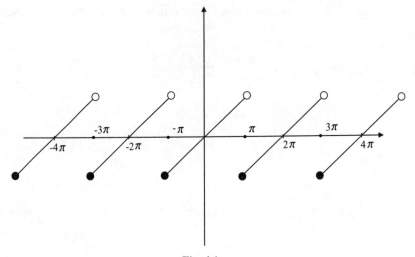

Fig. 4.1.

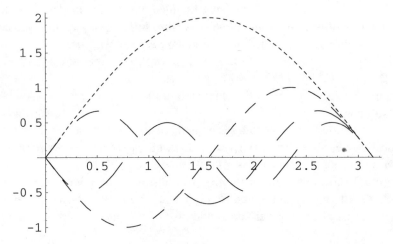

Fig. 4.2.

SOLUTION
Following our formulas, we calculate

$$a_0 = \frac{1}{\pi} \int_{-\pi}^{\pi} f(x)\,dx = \frac{1}{\pi} \int_{-\pi}^{0} 0\,dx + \frac{1}{\pi} \int_{0}^{\pi} \pi\,dx = \pi.$$

$$a_n = \frac{1}{\pi} \int_{0}^{\pi} \pi \cos nx\,dx = 0, \quad \text{all } n \geq 1.$$

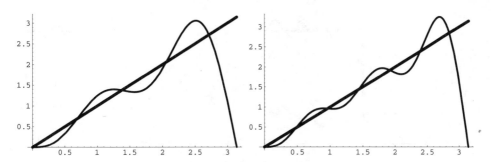

Fig. 4.3.   The sum of four terms and of six terms of the Fourier series.

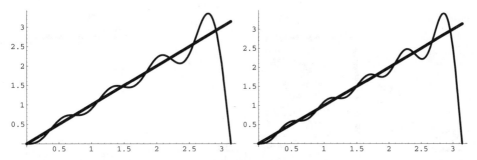

Fig. 4.4.   The sum of eight terms and of ten terms of the Fourier series.

$$b_n = \frac{1}{\pi} \int_0^\pi \pi \sin nx \, dx = \frac{1}{n}(1 - \cos n\pi) = \frac{1}{n}\left[1 - (-1)^n\right].$$

Another way to write this last calculation is

$$b_{2n} = 0, \quad b_{2n-1} = \frac{2}{2n-1}.$$

In sum, the Fourier expansion for f is

$$f(x) = \frac{\pi}{2} + 2\left(\sin x + \frac{\sin 3x}{3} + \frac{\sin 5x}{5} + \cdots\right).$$

Figure 4.5 shows the fourth and sixth partial sums, compared against $f(x)$.
Figure 4.6 shows the eighth and tenth partial sums, compared against $f(x)$.

Math Note: The places on the graphs where the Fourier series deviates sharply
from the true function—usually at endpoints and corners—are of particular interest.
These places show up in music as unwanted noise and hiss. Filters are constructed
using Fourier analysis in order to remove these artifacts.

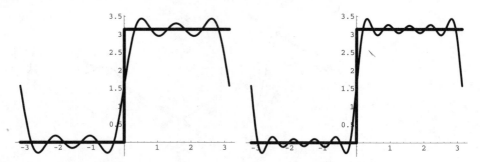

Fig. 4.5. The sum of four terms and of six terms of the Fourier series.

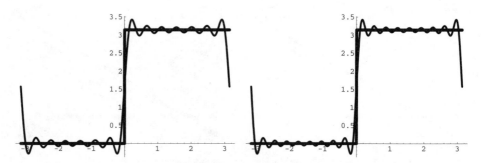

Fig. 4.6. The sum of eight terms and of ten terms of the Fourier series.

e.g. **EXAMPLE 4.3**

Find the Fourier series of the function given by

$$f(x) = \begin{cases} -\pi/2 & \text{if } -\pi \leq x < 0 \\ \pi/2 & \text{if } 0 \leq x \leq \pi. \end{cases}$$

SOLUTION

This is the same function as in the last example, with $\pi/2$ subtracted. Thus the Fourier series may be obtained by subtracting the quantity $\pi/2$ from the Fourier series that we obtained in that example. The result is

$$f(x) = 2\left(\sin x + \frac{\sin 3x}{3} + \frac{\sin 5x}{5} + \cdots\right).$$

The graph of this function, suitably periodized, is shown in Fig. 4.7.

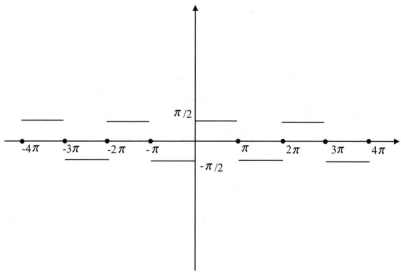

Fig. 4.7.   Graph of the function $f(x) = \begin{cases} -\pi/2 & \text{if } -\pi \le x < 0 \\ \pi/2 & \text{if } 0 \le x \le \pi. \end{cases}$

You Try It: Calculate the Fourier series of the function

$$f(x) = x^2 - x, \qquad 0 \le x < 2\pi.$$

EXAMPLE 4.4
Calculate the Fourier series of the function

$$f(x) = \begin{cases} -\pi/2 - x/2 & \text{if } -\pi \le x < 0 \\ \pi/2 - x/2 & \text{if } 0 \le x \le \pi. \end{cases}$$

SOLUTION
This function is simply the function from Example 4.3 minus half the function from Example 4.1. Thus we may obtain the requested Fourier series by subtracting half the series from Example 4.3 from the series in Example 4.1. The result is

$$f(x) = 2\left(\sin x + \frac{\sin 3x}{3} + \frac{\sin 5x}{5} + \cdots \right)$$

$$- \left(\sin x - \frac{\sin 2x}{2} + \frac{\sin 3x}{3} - + \cdots \right)$$

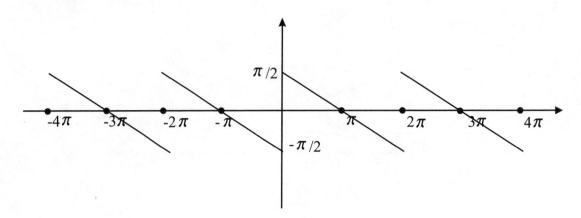

Fig. 4.8.

$$= \sin x + \frac{\sin 2x}{2} + \frac{\sin 3x}{3} + \cdots$$

$$= \sum_{n=1}^{\infty} \frac{\sin nx}{n}.$$

The graph of this series is the sawtooth wave shown in Fig. 4.8.

☞ **You Try It:**   Calculate the Fourier series of the function

$$f(x) = \begin{cases} -x & \text{if } -\pi \le x < 0 \\ x & \text{if } 0 \le x \le \pi. \end{cases}$$

4.2 Some Remarks About Convergence

The study of convergence of Fourier series is both deep and subtle. It would take us far afield to consider this matter in any detail. In the present section we shall very briefly describe a few of the basic results, but we shall not prove them.

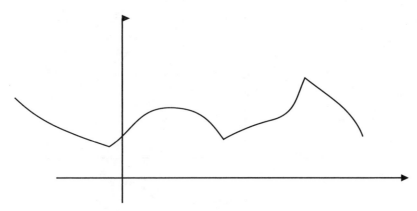

Fig. 4.9.

Our basic pointwise convergence result for Fourier series, which finds its genesis in work of Dirichlet (1805–1859), is this:

DEFINITION 4.1
Let f be a function on $[-\pi, \pi]$. We say that f is *piecewise smooth* if the graph of f consists of finitely many continuously differentiable curves, and furthermore that the one-sided derivatives exist at each of the endpoints $\{p_1, \ldots, p_k\}$ of the definition of the curves, in the sense that

$$\lim_{h \to 0^+} \frac{f(p_j + h) - f(p_j)}{h} \quad \text{and} \quad \lim_{h \to 0^-} \frac{f(p_j + h) - f(p_j)}{h}$$

exist. Further, we require that f' extend continuously to $[p_j, p_{j+1}]$ for each $j = 1, \ldots, k - 1$. See Fig. 4.9.

THEOREM 4.1   Given
Let f be a function on $[-\pi, \pi]$ which is piecewise smooth and overall continuous. Then the Fourier series of f converges at each point c of $[-\pi, \pi]$ to $f(c)$.

Let f be a function on the interval $[-\pi, \pi]$. We say that f has a *simple discontinuity* (or a *discontinuity of the first kind*) at the point $c \in (-\pi, \pi)$ if the limits $\lim_{x \to c^-} f(x)$ and $\lim_{x \to c^+} f(x)$ exist and

$$\lim_{x \to c^-} f(x) \neq \lim_{x \to c^+} f(x).$$

You should understand that a simple discontinuity is in contradistinction to the other kind of discontinuity. We say that f has a *discontinuity of the second kind* at c if either $\lim_{x \to c^-} f(x)$ or $\lim_{x \to c^-} f(x)$ does not exist.

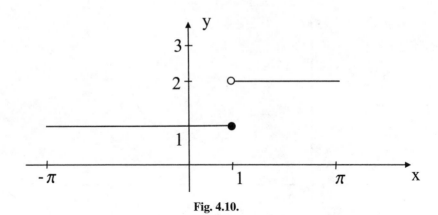

Fig. 4.10.

EXAMPLE 4.5
The function

$$f(x) = \begin{cases} 1 & \text{if } -\pi \le x \le 1 \\ 2 & \text{if } 1 < x \le \pi \end{cases}$$

has a simple discontinuity at $x = 1$. It is continuous at all other points of the interval $[-\pi, \pi]$. See Fig. 4.10.

The function

$$g(x) = \begin{cases} \sin\frac{1}{x} & \text{if } x \ne 0 \\ 0 & \text{if } x = 0 \end{cases}$$

has a discontinuity of the second kind at the origin. See Fig. 4.11.

Our next result about convergence is a bit more technical to state, but it is important in practice, and has historically been very influential. It is due to L. Fejér.

DEFINITION 4.2
Let f be a function and let

$$\frac{1}{2}a_0 + \sum_{n=1}^{\infty} (a_n \cos nx + b_n \sin nx)$$

be its Fourier series. The Nth *partial sum* of this series is

$$S_N(f)(x) = \frac{1}{2}a_0 + \sum_{n=1}^{N} (a_n \cos nx + b_n \sin nx).$$

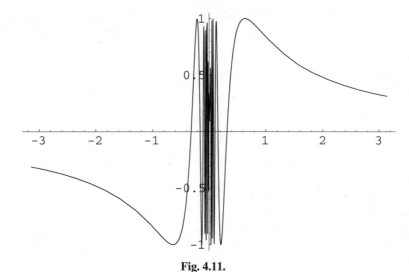

Fig. 4.11.

The *Cesàro mean* of the series is

$$\sigma_N(f)(x) = \frac{1}{N+1}\sum_{j=0}^{N} S_j(f)(x).$$

In other words, the Cesàro means are simply the averages of the partial sums.

THEOREM 4.2

Let f be a continuous function on the interval $[-\pi, \pi]$. Then the Cesàro means $\sigma_N(f)$ of the Fourier series for f converge uniformly to f.

A useful companion result is this:

THEOREM 4.3

Let f be a piecewise continuous function on $[-\pi, \pi]$—meaning that the graph of f consists of finitely many continuous curves. Let p be the endpoint of one of those curves, and assume that $\lim_{x \to p-} f(x) \equiv f(p^-)$ and $\lim_{x \to p+} f(x) \equiv f(p^+)$ exist. Then the Fourier series of f at p converges to $[f(p^-) + f(p^+)]/2$.

In fact, with a few more hypotheses, we may make the result even sharper. Recall that a function f is *monotone increasing* if $x_1 \leq x_2$ implies $f(x_1) \leq f(x_2)$. The function is *monotone decreasing* if $x_1 \leq x_2$ implies $f(x_1) \geq f(x_2)$. If the function is either monotone increasing or monotone decreasing, then we just call it *monotone*. Now we have this result of Dirichlet:

Given

THEOREM 4.4

Let f be a function on $[-\pi, \pi]$ which is piecewise continuous. Assume that each piece of f is monotone. Then the Fourier series of f converges at each point of continuity c of f in $[-\pi, \pi]$ to $f(c)$. At other points x it converges to $[f(x^-) + f(x^+)]/2$.

The hypotheses in this theorem are commonly referred to as the *Dirichlet conditions*.

By linearity, we may extend this last result to functions that are piecewise the difference of two monotone functions. Such functions are said to be of *bounded variation*, and exceed the scope of the present book. See [KRA2] for a detailed discussion. The book [TIT] discusses convergence of the Fourier series of such functions.

Math Note: A function f is said to be of bounded variation on an interval $[a, b]$ if there is a constant $C > 0$ such that, for each partition $\mathcal{P} = \{x_0, x_1, \ldots, x_k\}$ of the interval, it holds that

$$\sum_{j=1}^{k} |f(x_j) - f(x_{j-1})| \le C.$$

Such a function has a bound on its total oscillation. It can be shown that f is of bounded variation if and only if f can be written as the difference of two monotone functions.

4.3   Even and Odd Functions: Cosine and Sine Series

A function f is said to be *even* if $f(-x) = f(x)$. A function g is said to be *odd* if $g(-x) = -g(x)$.

e.g.

EXAMPLE 4.6

The function $f(x) = \cos x$ is even because $\cos(-x) = \cos x$. The function $g(x) = \sin x$ is odd because $\sin(-x) = -\sin x$.

   You Try It: Which of these functions is odd and which even?

$$f(x) = x \sin x, \ g(x) = x \cos x, \ h(x) = x^3, \ k(x) = \tan x, \ m(x) = e^{-x^2}.$$

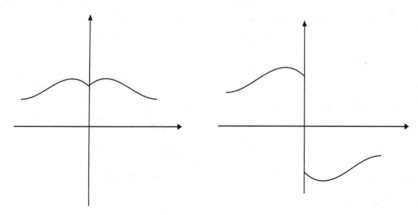

Fig. 4.12. An even and an odd function.

The graph of an even function is symmetric about the y-axis. The graph of an odd function is is skew-symmetric about the y-axis. Refer to Fig. 4.12.

If f is even on the interval $[-a, a]$, then

$$\int_{-a}^{a} f(x)\, dx = 2 \int_{0}^{a} f(x)\, dx \tag{1}$$

and if f is odd on the interval $[-a, a]$, then

$$\int_{-a}^{a} f(x)\, dx = 0. \tag{2}$$

Finally, we have the following parity relations

$$(\text{even}) \cdot (\text{even}) = (\text{even}) \qquad (\text{even}) \cdot (\text{odd}) = (\text{odd})$$

$$(\text{odd}) \cdot (\text{odd}) = (\text{even}).$$

Now suppose that f is an even function on the interval $[-\pi, \pi]$. Then $f(x) \cdot \sin nx$ is odd, and therefore

$$b_n = \frac{1}{\pi} \int_{-\pi}^{\pi} f(x) \sin nx\, dx = 0.$$

For the cosine coefficients, we have

$$a_n = \frac{1}{\pi} \int_{-\pi}^{\pi} f(x) \cos nx\, dx = \frac{2}{\pi} \int_{0}^{\pi} f(x) \cos nx\, dx.$$

Thus the Fourier series for an even function contains only cosine terms.

By the same token, suppose now that f is an odd function on the interval $[-\pi, \pi]$. Then $f(x) \cdot \cos nx$ is an odd function, and therefore

$$a_n = \frac{1}{\pi} \int_{-\pi}^{\pi} f(x) \cos nx \, dx = 0.$$

For the sine coefficients, we have

$$b_n = \frac{1}{\pi} \int_{-\pi}^{\pi} f(x) \sin nx \, dx = \frac{2}{\pi} \int_{0}^{\pi} f(x) \sin nx \, dx.$$

Thus the Fourier series for an odd function contains only sine terms.

EXAMPLE 4.7
Examine the Fourier series of the function $f(x) = x$ from the point of view of even/odd.

SOLUTION
The function is odd, so the Fourier series must be a sine series. We calculated in Example 4.1 that the Fourier series is in fact

$$x = f(x) = 2 \left(\sin x - \frac{\sin 2x}{2} + \frac{\sin 3x}{3} - + \cdots \right). \tag{3}$$

The expansion is valid on $(-\pi, \pi)$, but not at the endpoints (since the series of course sums to 0 at $-\pi$ and π).

EXAMPLE 4.8
Examine the Fourier series of the function $f(x) = |x|$ from the point of view of even/odd.

SOLUTION
The function is even, so the Fourier series must be a cosine series. In fact we see that

$$a_0 = \frac{1}{\pi} \int_{-\pi}^{\pi} |x| \, dx = \frac{2}{\pi} \int_{0}^{\pi} x \, dx = \pi.$$

Also, for $n \geq 1$,

$$a_n = \frac{2}{\pi} \int_{0}^{\pi} |x| \cos nx \, dx = \frac{2}{\pi} \int_{0}^{\pi} x \cos nx \, dx.$$

An integration by parts gives that

$$a_n = \frac{2}{\pi n^2} (\cos n\pi - 1) = \frac{2}{\pi n^2} [(-1)^n - 1].$$

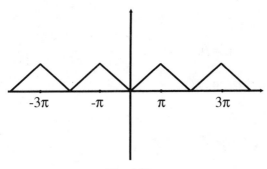

Fig. 4.13.

As a result,

$$a_{2j} = 0 \quad \text{and} \quad a_{2j-1} = -\frac{4}{\pi(2j-1)^2}.$$

In conclusion,

$$|x| = \frac{\pi}{2} - \frac{4}{\pi}\left(\cos x + \frac{\cos 3x}{3^2} + \frac{\cos 5x}{5^2} + \cdots\right). \qquad (4)$$

The periodic extension of the original function $f(x) = |x|$ on $[-\pi, \pi]$ is depicted in Fig. 4.13. By Theorem 4.1, the series converges to f at every point of $[-\pi, \pi]$.

It is worth noting that $x = |x|$ on $[0, \pi]$. Thus the expansions (3) and (4) represent the same function on that interval. Of course (3) is the *Fourier sine series* for x on $[0, \pi]$ while (4) is the *Fourier cosine series* for x on $[0, \pi]$. More generally, if g is *any* integrable function on $[0, \pi]$, we may take its odd extension $\tilde{g}$ to $[-\pi, \pi]$ and calculate the Fourier series. The result will be the Fourier sine series expansion for g on $[0, \pi]$. Instead we could take the even extension $\tilde{\tilde{g}}$ to $[-\pi, \pi]$ and calculate the Fourier series. The result will be the Fourier cosine series expansion for g on $[0, \pi]$.

EXAMPLE 4.9

Find the Fourier sine series and the Fourier cosine series expansions for the function $f(x) = \cos x$ on the interval $[0, \pi]$.

SOLUTION

Of course the Fourier series expansion of $\tilde{f}$ contains only sine terms. Its coefficients will be

$$b_n = \frac{2}{\pi}\int_0^\pi \cos x \sin nx \, dx = \begin{cases} 0 & \text{if } n = 1 \\ \dfrac{2n}{\pi}\left[\dfrac{1 + (-1)^n}{n^2 - 1}\right] & \text{if } n > 1. \end{cases}$$

As a result,

$$b_{2j-1} = 0 \qquad \text{and} \qquad b_{2j} = \frac{8j}{\pi(4j^2 - 1)}.$$

The sine series for f is therefore

$$\cos x = f(x) = \frac{8}{\pi} \sum_{j=1}^{\infty} \frac{j \sin 2jx}{4j^2 - 1}, \quad 0 < x < \pi.$$

To obtain the cosine series for f, we consider the even extension $\widetilde{\widetilde{f}}$. Of course all the b_n will vanish. Also

$$a_n = \frac{2}{\pi} \int_0^{\pi} \cos x \sin nx \, dx = \begin{cases} 1 & \text{if } n = 1 \\ 0 & \text{if } n > 1. \end{cases}$$

We therefore see, not surprisingly, that the Fourier cosine series for cosine on $[0, \pi]$ is the single summand $\cos x$.

 You Try It: Find the cosine series expansion for $f(x) = x - x^2$. Now find the sine series expansion for f.

 Math Note: You can use the idea of parity and reflection (and translation) to produce the sine series or the cosine series of a function on any interval with multiples of π as endpoints. As an example, calculate the cosine series of $f(x) = x$ on the interval $[3\pi, 4\pi]$.

4.4 Fourier Series on Arbitrary Intervals

We have developed Fourier analysis on the interval $[-\pi, \pi]$ (resp. the interval $[0, \pi]$) just because it is notationally convenient. In particular,

$$\int_{-\pi}^{\pi} \cos jx \cos kx \, dx = 0 \qquad \text{for } j \neq k$$

and so forth. This fact is special to the interval of length 2π. But many physical problems take place on an interval of some other length. We must therefore be able to adapt our analysis to intervals of any length. This amounts to a straightforward change of scale on the horizontal axis. We treat the matter in the present section.

Now we concentrate our attention on an interval of the form $[-L, L]$. As x runs from $-L$ to L, we will have a corresponding variable t that runs from $-\pi$ to π. We mediate between these two variables using the formulas

$$t = \frac{\pi x}{L} \qquad \text{and} \qquad x = \frac{Lt}{\pi}.$$

Thus the function $f(x)$ on $[-L, L]$ is transformed to a new function $\widetilde{f}(t) \equiv f(Lt/\pi)$ on $[-\pi, \pi]$.

If f satisfies the conditions for convergence of the Fourier series, then so will $\widetilde{f}$, and vice versa. Thus we may consider the Fourier expansion

$$\widetilde{f}(t) = \frac{1}{2}a_0 + \sum_{n=1}^{\infty} (a_n \cos nt + b_n \sin nt).$$

Here, of course,

$$a_n = \frac{1}{\pi} \int_{-\pi}^{\pi} \widetilde{f}(t) \cos nt \, dt \qquad \text{and} \qquad b_n = \frac{1}{\pi} \int_{-\pi}^{\pi} \widetilde{f}(t) \sin nt \, dt.$$

Now let us write out these last two formulas and perform a change of variables. We find that

$$a_n = \frac{1}{\pi} \int_{-\pi}^{\pi} f(Lt/\pi) \cos nt \, dt$$

$$= \frac{1}{\pi} \int_{-L}^{L} f(x) \cos \frac{n\pi x}{L} \cdot \frac{\pi}{L} \, dx$$

$$= \frac{1}{L} \int_{-L}^{L} f(x) \cos \frac{n\pi x}{L} \, dx.$$

Likewise,

$$b_n = \frac{1}{L} \int_{-L}^{L} f(x) \sin \frac{n\pi x}{L} \, dx.$$

EXAMPLE 4.10   e.g.

Calculate the Fourier series on the interval $[-2, 2]$ of the function

$$f(x) = \begin{cases} 0 & \text{if } -2 \le x < 0 \\ 1 & \text{if } 0 \le x \le 2. \end{cases}$$

SOLUTION

Of course $L = 2$, so we calculate that

$$a_n = \frac{1}{2} \int_0^2 \cos \frac{n\pi x}{2} \, dx = \begin{cases} 1 & \text{if } n = 0 \\ 0 & \text{if } n \geq 1. \end{cases}$$

Also

$$b_n = \frac{1}{2} \int_0^2 \sin \frac{n\pi x}{L} \, dx = \frac{1}{n\pi}[1 - (-1)^n].$$

This may be rewritten as

$$b_{2j} = 0 \qquad \text{and} \qquad b_{2j-1} = \frac{2}{(2j-1)\pi}.$$

In conclusion,

$$f(x) = g(t) = \frac{1}{2}a_0 + \sum_{n=1}^{\infty} (a_n \cos nt + b_n \sin nt)$$

$$= \frac{1}{2} + \sum_{j=1}^{\infty} \frac{2}{(2j-1)\pi} \sin\left[(2j-1) \cdot \frac{\pi x}{2}\right].$$

EXAMPLE 4.11

Calculate the Fourier series of the function $f(x) = \cos x$ on the interval $[-\pi/2, \pi/2]$.

SOLUTION

We calculate that

$$a_0 = \frac{2}{\pi} \int_{-\pi/2}^{\pi/2} \cos x \cdot \cos x \, dx = 1.$$

Also, for $n \geq 1$,

$$a_n = \frac{1}{\pi} \int_{-\pi/2}^{\pi/2} \cos x \cos nx \, dx$$

$$= \frac{2}{\pi} \int_{-\pi/2}^{\pi/2} \frac{1}{2} [\cos(n+1)x + \cos(n-1)x] \, dx$$

$$= \frac{1}{\pi} \left(\frac{\sin(n+1)x}{n+1} + \frac{\sin(n+1)x}{n+1} \right) \Bigg|_{-\pi/2}^{\pi/2}$$

$$= \begin{cases} 0 & \text{if n is odd} \\ \dfrac{2}{\pi(n^2-1)} & \text{if } n = 2m, n \neq 4m \\ \dfrac{-2}{\pi(n^2-1)} & \text{if } n = 4m. \end{cases}$$

A similar calculation shows that

$$b_n = \frac{1}{\pi}\int_{-\pi/2}^{\pi/2} \cos x \sin nx \, dx$$

$$= \frac{2}{\pi}\int_{-\pi/2}^{\pi/2} \frac{1}{2}[\sin(n+1)x + \sin(n-1)x] \, dx$$

$$= \frac{1}{\pi}\left(\frac{-\cos(n+1)x}{n+1} + \frac{-\cos(n+1)x}{n+1} \right)\Bigg|_{-\pi/2}^{\pi/2}$$

$$= \begin{cases} 0 & \text{if n is even} \\ \dfrac{2}{\pi(n^2-1)} & \text{if } n = 2m+1, n \neq 4m+1 \\ \dfrac{-2}{\pi(n^2-1)} & \text{if } n = 4m+1. \end{cases}$$

As a result, the Fourier series expansion for $\cos x$ on the interval $[-\pi/2, \pi/2]$ is

$$\cos x = f(x)$$

$$= \frac{1}{2} + \sum_{j=1}^{\infty} \frac{2}{\pi([2(2j-1)]^2-1)} \cos([2(2j-1)]2nx)$$

$$+ \sum_{j=1}^{\infty} \frac{-2}{\pi([4j]^2-1)} \cos([4j]2nx)$$

$$+ \sum_{j=1}^{\infty} \frac{2}{\pi([2(2j-1)+1]^2-1)} \sin([2(2j-1)+1]2nx)$$

$$+ \sum_{j=1}^{\infty} \frac{2}{\pi([4j+1]^2-1)} \sin([(4j+1)+1]2nx).$$

You Try It: Find the Fourier series expansion for $f(x) = x - x^2$ on the interval $[-1, 1]$.

Math Note: We can combine the ideas of the present section with those of the last section to produce the Fourier sine series or cosine series of a function on any interval centered about the origin. Implement this thought to calculate the cosine series of $g(x) = x^2 - x$ on the interval $[0, 2]$.

4.5 Orthogonal Functions

In the classical Euclidean geometry of 3-space, just as we learn in multivariable calculus class, one of the key ideas is that of orthogonality. Let us briefly review it now.

If $\mathbf{v} = \langle v_1, v_2, v_3 \rangle$ and $\mathbf{w} = \langle w_1, w_2, w_3 \rangle$ are vectors in $\mathbb{R}^3$, then we define their *dot product*, or *inner product*, or *scalar product* to be

$$\mathbf{v} \cdot \mathbf{w} = v_1 w_1 + v_2 w_2 + v_3 w_3.$$

What is the interest of the inner product? There are three answers:

- Two vectors are perpendicular or *orthogonal*, written $\mathbf{v} \perp \mathbf{w}$, if and only if $\mathbf{v} \cdot \mathbf{w} = 0$.
- The *length* of a vector is given by

$$\|\mathbf{v}\| = \sqrt{\mathbf{v} \cdot \mathbf{v}}.$$

- The *angle* θ between two vectors $\mathbf{v}$ and $\mathbf{w}$ is given by

$$\cos \theta = \frac{\mathbf{v} \cdot \mathbf{w}}{\|\mathbf{v}\| \|\mathbf{w}\|}.$$

In fact all of the geometry of 3-space is built on these three facts.

One of the great ideas of twentieth-century mathematics is that many other spaces—sometimes abstract spaces, and sometimes infinite-dimensional spaces—can be equipped with an inner product that endows that space with a useful geometry. That is the idea that we shall explore in the present section.

Let X be a vector space. This means that X is equipped with (i) a notion of addition and (ii) a notion of scalar multiplication. These two operations are hypothesized to satisfy the expected properties: addition is commutative and associative, scalar multiplication is commutative, associative, and distributive, and so forth. We say that X is equipped with an *inner product* (which we now denote by $\langle\,,\,\rangle$) if there is a binary operation

$$\langle\,,\,\rangle : X \times X \to \mathbb{R}$$

satisfying

1. $\langle \mathbf{u} + \mathbf{v}, \mathbf{w} \rangle = \langle \mathbf{u}, \mathbf{w} \rangle$;
2. $\langle c\mathbf{u}, \mathbf{v} \rangle = c\langle \mathbf{u}, \mathbf{v} \rangle$;
3. $\langle \mathbf{u}, \mathbf{u} \rangle \geq 0$ and $\langle \mathbf{u}, \mathbf{u} \rangle = 0$ if and only if $\mathbf{u} = 0$;
4. $\langle \mathbf{u}, \mathbf{v} \rangle = \langle \mathbf{v}, \mathbf{u} \rangle$.

We shall give some interesting examples of inner products below. Before we do, let us note that an inner product as just defined gives rise to a notion of length, or a *norm*. Namely, we define

$$\|\mathbf{v}\| = \sqrt{\langle \mathbf{v}, \mathbf{v} \rangle}.$$

By property (3), we see that $\|\mathbf{v}\| \geq 0$ and $\|\mathbf{v}\| = 0$ if and only if $\mathbf{v} = 0$.

In fact the two key properties of the inner product and the norm are enunciated in the following proposition:

PROPOSITION 4.1
Let X be a vector space and $\langle\,,\,\rangle$ an inner product on that space. Let $\|\ \ \|$ be the induced norm. Then

(1) **The Cauchy–Schwarz–Bunjakovski inequality:** *If $\mathbf{u}, \mathbf{v} \in X$, then*

$$|\mathbf{u} \cdot \mathbf{v}| \leq \|\mathbf{u}\| \cdot \|\mathbf{v}\|.$$

(2) **The Triangle inequality:** *If $\mathbf{u}, \mathbf{v} \in X$, then*

$$\|\mathbf{u} + \mathbf{v}\| \leq \|\mathbf{u}\| + \|\mathbf{v}\|.$$

Just as an exercise, we shall derive the Triangle inequality from the Cauchy–Schwarz–Bunjakovski inequality. We have

$$\|\mathbf{u} + \mathbf{v}\|^2 = \langle (\mathbf{u} + \mathbf{v}), (\mathbf{u} + \mathbf{v}) \rangle$$

$$= \langle \mathbf{u}, \mathbf{u} \rangle + \langle \mathbf{u}, \mathbf{v} \rangle + \langle \mathbf{v}, \mathbf{u} \rangle + \langle \mathbf{v}, \mathbf{v} \rangle$$

$$= \|\mathbf{u}\|^2 + \|\mathbf{v}\|^2 + 2\langle \mathbf{u}, \mathbf{v} \rangle$$

$$\leq \|\mathbf{u}\|^2 + \|\mathbf{v}\|^2 + 2\|\mathbf{u}\| \cdot \|\mathbf{v}\|$$

$$= (\|\mathbf{u}\| + \|\mathbf{v}\|)^2.$$

Now taking the square root of both sides completes the argument.

EXAMPLE 4.12
Let $X = C[0, 1]$, the continuous functions on the interval $[0, 1]$. This is certainly a vector space with the usual notions of addition of functions and scalar multiplication of functions. We define an inner product by

$$\langle f, g \rangle = \int_0^1 f(x)g(x)\,dx$$

for any $f, g \in X$.

Then it is straightforward to verify that this definition of inner product satisfies all our axioms. Thus we may *define* two functions to be orthogonal if

$$\langle f, g \rangle = 0.$$

We say that the angle θ between two functions is given by

$$\cos \theta = \frac{\langle f, g \rangle}{\|f\|\|g\|}.$$

The *length* or *norm* of an element $f \in X$ is given by

$$\|f\| = \sqrt{\langle f, f \rangle} = \left[\int_0^1 f(x)^2\,dx \right]^{1/2}.$$

EXAMPLE 4.13
Let X be the space of all sequences $\{a_j\}$ with the property that $\sum_{j=1}^{\infty} |a_j|^2 < \infty$. This is a vector space with the obvious notions of addition and scalar multiplication. Define an inner product by

$$\langle \{a_j\}, \{b_j\} \rangle = \sum_{j=1}^{\infty} a_j b_j.$$

Then this inner product satisfies all our axioms.

For the purposes of studying Fourier series, the most important inner product space is that which we call $L^2[-\pi, \pi]$. This is the space of functions f on the interval $[-\pi, \pi]$ with the property that

$$\int_{-\pi}^{\pi} f(x)^2\,dx < \infty.$$

The inner product on this space is

$$\langle f, g \rangle = \int_{-\pi}^{\pi} f(x)g(x)\,dx.$$

One must note here that, by a variant of the Cauchy–Schwarz–Bunjakovski inequality, it holds that if $f, g \in L^2$ then the integral $\int f \cdot g \, dx$ exists and is finite. So our inner product makes sense.

Math Note: In fact if $w \geq 0$ is any weight function then we may define the space $L^2(w)$ to be the collection of all functions on the interval $[-\pi, \pi]$ that satisfy the condition

$$\int_{-\pi}^{\pi} f(x)^2 w(x) \, dx < \infty.$$

This type of weighted function space has become very important in modern analysis. Of course the inner product on this space is

$$\langle f, g \rangle = \int_{-\pi}^{\pi} f(x)g(x)w(x) \, dx.$$

This inner product will satisfy both the Cauchy–Schwarz–Bunjakovski and the Triangle inequalities.

Exercises

1. Find the Fourier series of the function

$$f(x) = \begin{cases} \pi & \text{if } -\pi \leq x \leq \dfrac{\pi}{2} \\ 0 & \text{if } \dfrac{\pi}{2} < x \leq \pi. \end{cases}$$

2. Find the Fourier series for the function

$$f(x) = \begin{cases} 0 & \text{if } -\pi \leq x < 0 \\ 1 & \text{if } 0 \leq x \leq \dfrac{\pi}{2} \\ 0 & \text{if } \dfrac{\pi}{2} < x \leq \pi. \end{cases}$$

3. Find the Fourier series of the function

$$f(x) = \begin{cases} 0 & \text{if } -\pi \leq x < 0 \\ \sin x & \text{if } 0 \leq x \leq \pi. \end{cases}$$

4. Find the Fourier series for the periodic function defined by

$$f(x) = \begin{cases} -\pi & \text{if } -\pi \leq x < 0 \\ x & \text{if } 0 \leq x \leq \pi. \end{cases}$$

5. Show that the Fourier series for the periodic function

$$f(x) = \begin{cases} 0 & \text{if } -\pi \le x < 0 \\ x^2 & \text{if } 0 \le x \le \pi \end{cases}$$

is

$$f(x) = \frac{\pi^2}{6} + 2\sum_{j=1}^{\infty}(-1)^j \frac{\cos jx}{j^2}$$

$$+ \pi \sum_{j=1}^{\infty}(-1)^{j+1}\frac{\sin jx}{j} - \frac{4}{\pi}\sum_{j=1}^{\infty}\frac{\sin(2j-1)x}{(2j-1)^3}.$$

6. Determine whether each of the following functions is even, odd, or neither:

$$x^5 \sin x, \quad e^x, \quad (\sin x)^3, \quad \sin x^2, \quad x + x^2 + x^3, \quad \ln \frac{1+x}{1-x}.$$

7. Show that any function f defined on a symmetrically placed interval can be written as the sum of an even function and an odd function. *Hint:* $f(x) = \frac{1}{2}[f(x) + f(-x)] + \frac{1}{2}[f(x) - f(-x)].$

8. Show that the sine series of the constant function $f(x) \equiv \pi/4$ is

$$\frac{\pi}{4} = \sin x + \frac{\sin 3x}{3} + \frac{\sin 5x}{5} + \cdots$$

for $0 < x < \pi$. What sum is obtained by setting $x = \pi/2$? What is the cosine series of this function?

9. Find the sine and the cosine series for $f(x) = \sin x$.

10. Find the Fourier series for these functions:

(a) $f(x) = \begin{cases} 1 + x & \text{if } -1 \le x < 0 \\ 1 - x & \text{if } 0 \le x \le 1 \end{cases}$

(b) $f(x) = |x|, \quad -2 \le x \le 2$

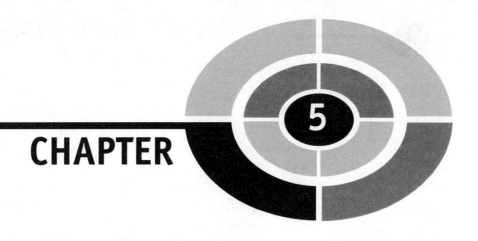

CHAPTER 5

Partial Differential Equations and Boundary Value Problems

5.1 Introduction and Historical Remarks

In the middle of the eighteenth century much attention was given to the problem of determining the mathematical laws governing the motion of a vibrating string with fixed endpoints at 0 and π (Fig. 5.1). An elementary analysis of tension shows that if $y(x, t)$ denotes the ordinate of the string at time t above the point x, then $y(x, t)$

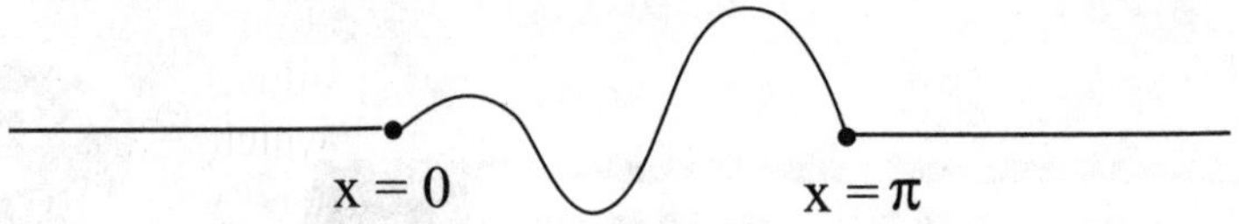

Fig. 5.1.

satisfies the *wave equation*

$$\frac{\partial^2 y}{\partial t^2} = a^2 \frac{\partial^2 y}{\partial x^2}.$$

Here a is a parameter that depends on the tension of the string. A change of scale will allow us to assume that $a = 1$. [A bit later we shall actually provide a formal derivation of the wave equation. See also [KRA3] for a more thorough consideration of these matters.]

In 1747 d'Alembert showed that solutions of this equation have the form

$$y(x, t) = \tfrac{1}{2}\left[f(at + x) + g(at - x)\right], \tag{1}$$

where f and g are "any" functions of one variable. [The following technicality must be noted: the functions f and g are initially specified on the interval $[0, \pi]$. We extend f and g to $[-\pi, 0]$ and to $[\pi, 2\pi]$ by odd reflection. Continue f and g to the rest of the real line so that they are 2π-periodic.]

In fact the wave equation, when placed in a "well-posed" setting, comes equipped with two boundary conditions:

$$y(x, 0) = \phi(x)$$

$$\partial_t y(x, 0) = \psi(x).$$

These conditions mean (i) that the wave has an initial configuration that is the graph of the function ϕ and (ii) that the string is released with initial velocity ψ.

If (1) is to be a solution of this boundary value problem, then f and g must satisfy

$$\tfrac{1}{2}\left[f(x) + g(-x)\right] = \phi(x) \tag{2}$$

and

$$\tfrac{1}{2}\left[f'(x) + g'(-x)\right] = \psi(x). \tag{3}$$

Integration of (3) gives a formula for $f(x) - g(-x)$. That and (2) give a system that may be solved for f and g with elementary algebra.

The converse statement holds as well: for any functions f and g, a function y of the form (1) satisfies the wave equation (check this as an exercise). The work of d'Alembert brought to the fore a controversy which had been implicit in the work of Daniel Bernoulli, Leonhard Euler, and others: what is a "function"? [We recommend the article [LUZ] for an authoritative discussion of the controversies that grew out of classical studies of the wave equation. See also [LAN].]

It is clear, for instance, in Euler's writings that he did not perceive a function to be an arbitrary "rule" that assigns points of the domain to points of the range; in particular, Euler did not think that a function could be specified in a fairly arbitrary fashion at different points of the domain. Once a function was specified on some small interval, Euler thought that it could only be extended in one way to a larger interval. Therefore, on physical grounds, Euler objected to d'Alembert's work. Euler's physical intuition ran contrary to his mathematical intuition. He claimed that the initial position of the vibrating string could be specified by several different functions pieced together continuously, so that a single f could not generate the motion of the string.

Daniel Bernoulli solved the wave equation by a different method (separation of variables, which we treat below) and was able to show that there are infinitely many solutions of the wave equation having the form

$$\phi_j(x, t) = \sin jx \cos jt.$$

Proceeding formally, he posited that all solutions of the wave equation satisfying $y(0, t) = y(\pi, t) = 0$ and $\partial_t y(x, 0) = 0$ will have the form

$$y = \sum_{j=1}^{\infty} a_j \sin jx \cos jt.$$

Setting $t = 0$ indicates that the initial form of the string is $f(x) \equiv \sum_{j=1}^{\infty} a_j \sin jx$. In d'Alembert's language, the initial form of the string is $\frac{1}{2}(f(x) - f(-x))$, for we know that

$$0 \equiv y(0, t) = f(t) + g(t)$$

(because the endpoints of the string are held stationary), hence $g(t) = -f(-t)$. If we suppose that d'Alembert's function is odd (as is $\sin jx$, each j), then the initial position is given by $f(x)$. Thus the problem of reconciling Bernoulli's solution to d'Alembert's reduces to the question of whether an "arbitrary" function f on $[0, \pi]$ may be written in the form $\sum_{j=1}^{\infty} a_j \sin jx$.

Since most mathematicians contemporary with Bernoulli believed that properties such as continuity, differentiability, and periodicity were preserved under (even infinite) addition, the consensus was that arbitrary f could *not* be represented as a (even infinite) trigonometric sum. The controversy extended over some years and was fueled by further discoveries (such as Lagrange's technique for interpolation by trigonometric polynomials) and more speculations.

In the 1820s, the problem of representation of an "arbitrary" function by trigonometric series was given a satisfactory answer as a result of two events. First there is the sequence of papers by Joseph Fourier culminating with the tract [FOU]. Fourier gave a formal method of expanding an "arbitrary" function f into a trigonometric series. He computed some partial sums for some sample f's and verified that they gave very good approximations to f. Secondly, Dirichlet proved the first theorem giving sufficient (and very general) conditions for the Fourier series of a function f to converge pointwise to f. *Dirichlet was one of the first, in 1828, to formalize the notions of partial sum and convergence of a series*; his ideas certainly had antecedents in work of Gauss and Cauchy.

For all practical purposes, these events mark the beginning of the mathematical theory of Fourier series (see [LAN]).

5.2 Eigenvalues, Eigenfunctions, and the Vibrating String

5.2.1 BOUNDARY VALUE PROBLEMS

We wish to motivate the physics of the vibrating string. We begin this discussion by seeking a nontrivial solution y of the differential equation

$$y'' + \lambda y = 0 \tag{1}$$

subject to the conditions

$$y(0) = 0 \qquad \text{and} \qquad y(\pi) = 0. \tag{2}$$

Notice that this is a different situation from the one we have studied in earlier parts of the book. In Chapter 2, on second-order linear equations, we usually had *initial conditions* $y(x_0) = y_0$ and $y'(x_0) = y_1$. Now we have what are called *boundary conditions*: we specify the value (*not* the derivative) of our solution at two different points. For instance, in the discussion of the vibrating string in the last section, we wanted our string to be pinned down at the two endpoints. These are typical boundary conditions coming from a physical problem.

The situation with boundary conditions is quite different from that for initial conditions. The latter is a sophisticated variation of the fundamental theorem of calculus. The former is rather more subtle. So let us begin to analyze.

First, if $\lambda < 0$, then it is known that any solution of (1) has at most one zero. So it certainly cannot satisfy the boundary conditions (2). Alternatively, we could just solve the equation explicitly when $\lambda < 0$ and see that the independent solutions are a pair of exponentials, no linear combination of which can satisfy (2).

If $\lambda = 0$, then the general solution of (1) is the linear function $y = Ax + B$. Such a function cannot vanish at two points unless it is identically zero.

So the only interesting case is $\lambda > 0$. In this situation, the general solution of (1) is

$$y = A \sin \sqrt{\lambda} x + B \cos \sqrt{\lambda} x.$$

Since $y(0) = 0$, this in fact reduces to

$$y = A \sin \sqrt{\lambda} x.$$

In order for $y(\pi) = 0$, we must have $\sqrt{\lambda}\pi = n\pi$ for some positive integer n, thus $\lambda = n^2$. These values of λ are termed the *eigenvalues* of the problem, and the corresponding solutions

$$\sin x, \quad \sin 2x, \quad \sin 3x, \ldots$$

are called the *eigenfunctions* of the problem (1), (2).

We note these immediate properties of the eigenvalues and eigenfunctions for our problem:

(i)   If ϕ is an eigenfunction for eigenvalue λ, then so is $c \cdot \phi$ for any constant c.

(ii) The eigenvalues $1, 4, 9, \ldots$ form an increasing sequence that approaches $+\infty$.

(iii) The nth eigenfunction $\sin nx$ vanishes at the endpoints $0, \pi$ (as we originally mandated) and has exactly $n - 1$ zeros in the interval $(0, \pi)$.

5.2.2 DERIVATION OF THE WAVE EQUATION

Now let us re-examine the vibrating string from the last section and see how eigenfunctions and eigenvalues arise naturally in this physical problem. We consider a flexible string with negligible weight that is fixed at its ends at the points $(0, 0)$ and $(\pi, 0)$. The curve is deformed into an initial position $y = f(x)$ in the x–y plane and then released.

Our analysis will ignore damping effects, such as air resistance. We assume that, in its relaxed position, the string is as in Fig. 5.2. The string is plucked in the vertical direction, and is thus set in motion in a vertical plane.

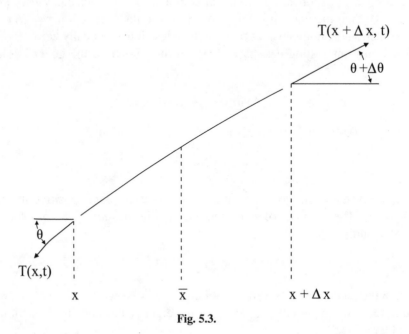

$$x = 0 \qquad x = \pi$$

Fig. 5.2.

$$T(x,t) \qquad \overline{x} \qquad x + \Delta x$$

Fig. 5.3.

We focus attention on an "element" Δx of the string (Fig. 5.3) that lies between x and $x + \Delta x$. We adopt the usual physical conceit of assuming that the displacement (motion) of this string element is *small*, so that there is only a slight error in supposing that the motion of each point of the string element is strictly vertical. We let the tension of the string, at the point x at time t, be denoted by $T(x, t)$. Note that T acts only in the tangential direction (i.e., along the string). We denote the mass density of the string by ρ.

Since *there is no horizontal component of acceleration*, we see that

$$T(x + \Delta x, t) \cdot \cos(\theta + \Delta\theta) - T(x, t) \cdot \cos(\theta) = 0. \tag{3}$$

[Refer to Fig. 5.4: The expression $T(\star) \cdot \cos(\star)$ denotes $H(\star)$, the horizontal component of the tension.] Thus equation (3) says that H is independent of x.

Now we look at the vertical component of force (acceleration):

$$T(x + \Delta x, t) \cdot \sin(\theta + \Delta\theta) - T(x, t) \cdot \sin(\theta) = \rho \cdot \Delta x \cdot u_{tt}(\overline{x}, t). \tag{4}$$

Here $\overline{x}$ is the mass center of the string element and we are applying Newton's second law—that the external force is the mass of the string element times the

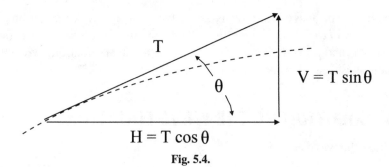

Fig. 5.4.

acceleration of its center of mass. We use subscripts to denote derivatives. We denote the vertical component of $T(\star)$ by $V(\star)$. Thus equation (4) can be written as

$$\frac{V(x + \Delta x, t) - V(x, t)}{\Delta x} = \rho \cdot u_{tt}(x, t).$$

Letting $\Delta x \to 0$ yields

$$V_x(x, t) = \rho \cdot u_{tt}(x, t). \tag{5}$$

We would like to express equation (5) entirely in terms of u, so we notice that

$$V(x, t) = H(t) \tan \theta = H(t) \cdot u_x(x, t).$$

[We have used the fact that the derivative in x is the slope of the tangent line, which is $\tan \theta$.] Substituting this expression for V into (5) yields

$$(H u_x)_x = \rho \cdot u_{tt}.$$

But H is independent of x, so this last line simplifies to

$$H \cdot u_{xx} = \rho \cdot u_{tt}.$$

For small displacements of the string, θ is nearly zero, so $H = T \cos \theta$ is nearly T. Thus we finally write our equation as

$$\frac{T}{\rho} u_{xx} = u_{tt}.$$

It is traditional to denote the constant on the left by a^2. We finally arrive at the *wave equation*

$$a^2 u_{xx} = u_{tt}.$$

Math Note: The wave equation is an instance of a class of equations called *hyperbolic partial differential equations*. There are also *elliptic equations* (such as the Laplacian) and *parabolic equations* (such as the heat equation). We shall say more about these as the book develops.

5.2.3 SOLUTION OF THE WAVE EQUATION

We consider the wave equation

$$a^2 y_{xx} = y_{tt} \tag{6}$$

with the boundary conditions

$$y(0, t) = 0$$

and

$$y(\pi, t) = 0.$$

Physical considerations dictate that we also impose the initial conditions

$$\left. \frac{\partial y}{\partial t} \right|_{t=0} = 0 \tag{7}$$

(indicating that the initial velocity of the string is 0) and

$$y(x, 0) = f(x) \tag{8}$$

(indicating that the initial configuration of the string is the graph of the function f).

We solve the wave equation using a classical technique known as "separation of variables." For convenience, we assume that the constant $a = 1$. We guess a solution of the form $u(x, t) = u(x) \cdot v(t)$. Putting this guess into the differential equation

$$u_{xx} = u_{tt}$$

gives

$$u''(x)v(t) = u(x)v''(t).$$

We may obviously separate variables, in the sense that we may write

$$\frac{u''(x)}{u(x)} = \frac{v''(t)}{v(t)}.$$

The left-hand side depends only on x while the right-hand side depends only on t. The only way this can be true is if

$$\frac{u''(x)}{u(x)} = \lambda = \frac{v''(t)}{v(t)}$$

for some constant λ. But this gives rise to two second-order linear, ordinary differential equations that we can solve explicitly:

$$u'' = \lambda \cdot u \tag{9}$$

$$v'' = \lambda \cdot v. \tag{10}$$

Observe that this is the *same* constant λ in both of these equations. Now, as we have already discussed, we want the initial configuration of the string to pass through the points $(0, 0)$ and $(\pi, 0)$. We can achieve these conditions by solving (9) with $u(0) = 0$ and $u(\pi) = 0$. But of course this is the eigenvalue problem that we treated at the beginning of the section. The problem has a nontrivial solution if and only if $\lambda = n^2$ for some positive integer n, and the corresponding eigenfunction is

$$u_n(x) = \sin nx.$$

For this same λ, the general solution of (10) is

$$v(t) = A \sin nt + B \cos nt.$$

If we impose the requirement that $v'(0) = 0$, so that (7) is satisfied, then $A = 0$ and we find the solution

$$v(t) = B \cos nt.$$

This means that the solution we have found of our differential equation with boundary and initial conditions is

$$y_n(x, t) = \sin nx \cos nt. \tag{11}$$

And in fact any finite sum with coefficients (or *linear combination*) of these solutions will also be a solution:

$$y = \alpha_1 \sin x \cos t + \alpha_2 \sin 2x \cos 2t + \cdots \alpha_k \sin kx \cos kt.$$

Ignoring the rather delicate issue of convergence (which was discussed a bit in Section 4.2), we may claim that any *infinite* linear combination of the solutions (11) will also be a solution:

$$y = \sum_{j=1}^{\infty} b_j \sin jx \cos jt. \tag{12}$$

Now we must examine the final condition (8). The mandate $y(x, 0) = f(x)$ translates to

$$\sum_{j=1}^{\infty} b_j \sin jx = y(x, 0) = f(x) \tag{13}$$

or

$$\sum_{j=1}^{\infty} b_j u_j(x) = y(x, 0) = f(x). \tag{14}$$

Thus we demand that f have a valid Fourier series expansion. We know from our studies in Chapter 4 that such an expansion is valid for a rather broad class of functions f. Thus the wave equation is solvable in considerable generality.

Now fix $m \neq n$. We know that our eigenfunctions u_j satisfy

$$u_m'' = -m^2 u_m \qquad \text{and} \qquad u_n'' = -n^2 u_n.$$

Multiply the first equation by u_n and the second by u_m and subtract. The result is

$$u_n u_m'' - u_m u_n'' = (n^2 - m^2) u_n u_m$$

or

$$[u_n u_m' - u_m u_n']' = (n^2 - m^2) u_n u_m.$$

We integrate both sides of this last equation from 0 to π and use the fact that $u_j(0) = u_j(\pi) = 0$ for every j. The result is

$$0 = [u_n u_m' - u_m u_n']\Big|_0^{\pi} = (n^2 - m^2) \int_0^{\pi} u_m(x) u_n(x)\, dx.$$

Thus

$$\int_0^{\pi} \sin mx \sin nx\, dx = 0 \qquad \text{for } n \neq m \tag{15}$$

or

$$\int_0^{\pi} u_m(x) u_n(x)\, dx = 0 \qquad \text{for } n \neq m. \tag{16}$$

Of course this is a standard fact from calculus. But now we understand it as an orthogonality condition, and we see how the condition arises naturally from the differential equation. A little later, we shall fit this phenomenon into the general context of Sturm–Liouville problems.

In view of the orthogonality condition (16), it is natural to integrate both sides of (14) against $u_k(x)$. The result is

$$\int_0^\pi f(x) \cdot u_k(x)\, dx = \int_0^\pi \left[\sum_{j=0}^\infty b_j u_j(x) \right] \cdot u_k(x)\, dx$$

$$= \sum_{j=0}^\infty b_j \int_0^\pi u_j(x) u_k(x)\, dx$$

$$= \frac{\pi}{2} b_k.$$

The b_k are the Fourier coefficients that we studied in Chapter 4.

Math Note: The calculations that we performed in this section can be fitted into a much more general context. We shall give a taste of these ideas in Section 5.5. Certainly orthogonality, and orthogonal expansions, is one of the most pervasive ideas in modern analysis.

5.3 The Heat Equation: Fourier's Point of View

In [FOU], Fourier considered variants of the following basic question. Let there be given an insulated, homogeneous rod of length π with initial temperature at each $x \in [0, \pi]$ given by a function $f(x)$ (Fig. 5.5). Assume that the endpoints are held at temperature 0, and that the temperature of each cross-section is constant. The problem is to describe the temperature $u(x, t)$ of the point x in the rod at time t.

Let us now indicate the manner in which Fourier solved his problem. First, it is required to write a differential equation which u satisfies. We shall derive such an equation using three physical principles:

(1) The density of heat energy is proportional to the temperature u, hence the amount of heat energy in any interval $[a, b]$ of the rod is proportional to $\int_a^b u(x, t)\, dx$.

(2) **(Newton's Law of Cooling).**   The rate at which heat flows from a hot place to a cold one is proportional to the difference in temperature. The infinitesimal version of this statement is that the rate of heat flow across a point x (from left to right) is some negative constant times $\partial_x u(x, t)$.

(3) **(Conservation of Energy).**   Heat has no sources or sinks.

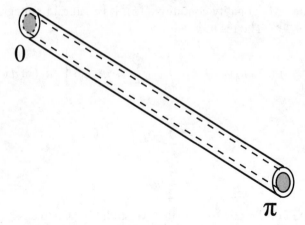

0

π

Fig. 5.5.

Now (3) tells us that the only way that heat can enter or leave any interval portion $[a, b]$ of the rod is through the endpoints. And (2) tells us exactly how this happens. Using (1), we may therefore write

$$\frac{d}{dt} \int_a^b u(x, t)\, dx = \eta^2 [\partial_x u(b, t) - \partial_x u(a, t)].$$

We may rewrite this equation as

$$\int_a^b \partial_t u(x, t)\, dx = \eta^2 \int_a^b \partial_x^2 u(x, t)\, dx.$$

Differentiating in b, we find that

$$\partial_t u = \eta^2 \partial_x^2 u, \tag{4}$$

and that is the heat equation.

Suppose for simplicity that the constant of proportionality η^2 equals 1. Fourier guessed that the equation (4) has a solution of the form $u(x, t) = \alpha(x)\beta(t)$. Substituting this guess into the equation yields

$$\alpha(x)\beta'(t) = \alpha''(x)\beta(t)$$

or

$$\frac{\beta'(t)}{\beta(t)} = \frac{\alpha''(x)}{\alpha(x)}.$$

Since the left side is independent of x and the right side is independent of t, it follows that there is a constant K such that

$$\frac{\beta'(t)}{\beta(t)} = K = \frac{\alpha''(x)}{\alpha(x)}$$

or

$$\beta'(t) = K\beta(t)$$

$$\alpha''(x) = K\alpha(x).$$

We conclude that $\beta(t) = Ce^{Kt}$. The nature of β, and hence of α, thus depends on the sign of K. But physical considerations tell us that the temperature will dissipate as time goes on, so we conclude that $K \leq 0$. Therefore $\alpha(x) = \cos\sqrt{-K}x$ and $\alpha(x) = \sin\sqrt{-K}x$ are solutions of the differential equation for α. The initial conditions $u(0, t) = u(\pi, t) = 0$ (since the ends of the rod are held at constant temperature 0) eliminate the first of these solutions and force $K = -j^2$, $j \in \mathbb{Z}$. Thus Fourier found the solutions

$$u_j(x, t) = e^{-j^2 t} \sin jx, \qquad j \in \mathbb{N}$$

of the heat equation. By linearity, any finite linear combination

$$\sum_j b_j e^{-j^2 t} \sin jx$$

of these solutions is also a solution. It is plausible to extend this assertion to infinite linear combinations. Using the initial condition $u(x, 0) = f(x)$ again raises the question of whether "any" function $f(x)$ on $[0, \pi]$ can be written as a (infinite) linear combination of the functions $\sin jx$.

Fourier's solution to this last problem (of the sine functions spanning essentially everything) is roughly as follows. Suppose f is a function that is so representable:

$$f(x) = \sum_j b_j \sin jx. \tag{5}$$

Setting $x = 0$ gives

$$f(0) = 0.$$

Differentiating both sides of (5) and setting $x = 0$ gives

$$f'(0) = \sum_{j=1}^{\infty} jb_j.$$

Successive differentiation of (5), and evaluation at 0, gives

$$f^{(k)}(0) = \sum_{j=1}^{\infty} j^k b_j (-1)^{[k/2]}.$$

for k odd (by oddness of f, the even derivatives must be 0 at 0). Here [ ] denotes the greatest integer function. Thus Fourier devised a system of infinitely many equations in the infinitely many unknowns $\{b_j\}$. He proceeded to solve this system by truncating it to an $N \times N$ system (the first N equations restricted to the first N unknowns), solved that truncated system, and then let N tend to ∞. Suffice it to say that Fourier's arguments contained many dubious steps (see [FOU] and [LAN]).

The upshot of Fourier's intricate and lengthy calculations was that

$$b_j = \frac{2}{\pi} \int_0^{\pi} f(x) \sin jx \, dx. \tag{6}$$

By modern standards, Fourier's reasoning was specious, for he began by assuming that f possessed an expansion in terms of sine functions. The formula (6) hinges on that supposition, together with steps in which one compensated division by zero with a later division by ∞. Nonetheless, Fourier's methods give an actual *procedure* for endeavoring to expand any given f in a series of sine functions.

Fourier's abstract arguments constitute the first part of his book. The bulk, and remainder, of the book consists of separate chapters in which the expansions for particular functions are computed.

Math Note: You will notice several parallels between our analysis of the heat equation in this section and the solution of the wave equation in Subsection 5.2.3. In both instances we assumed a solution of the form $\alpha(x)\beta(t)$. In both cases this led to trigonometric solutions. And for the general solution we considered a trigonometric series. Thus there are unifying principles that occur repeatedly in different parts of the theory of differential equations. Certainly Fourier series is one of those principles.

EXAMPLE 5.1

Suppose that the thin rod in the setup of the heat equation is first immersed in boiling water so that its temperature is uniformly 100°C. Then imagine that it is removed from the water at time $t = 0$ with its ends immediately put into ice so that these ends are kept at temperature 0°C. Find the temperature $u = u(x, t)$ under these circumstances.

SOLUTION

The initial temperature distribution is given by the constant function

$$f(x) = 100, \quad 0 < x < \pi.$$

The two boundary conditions, and the other initial condition, are as usual. Thus our job is simply this: to find the sine series expansion of this function f. Notice that $b_j = 0$ when j is even. For j odd, we calculate that

$$b_j = \frac{2}{\pi} \int_0^\pi 100 \sin jx \, dx = -\frac{200}{\pi} \frac{\cos jx}{j} \Big|_0^\pi$$

$$= \frac{400}{\pi j} \quad \text{as long as } j \text{ is odd.}$$

Thus

$$f(x) = \frac{400}{\pi} \left(\sin x + \frac{\sin 3x}{3} + \frac{\sin 5x}{5} + \cdots \right).$$

Now, referring to formula (5) from our general discussion of the heat equation, we know that

$$u(x, t) = \frac{400}{\pi} \left[e^{-a^2 t} \sin x + \frac{1}{3} e^{-9a^2 t} \sin 3x + \frac{1}{5} e^{-25a^2 t} \sin 5x + \cdots \right].$$

You Try It: If a rod of length 2 has its ends held steadily at temperatures 0°C and 100°C, then what is the steady-state temperature at the points of the rod?

EXAMPLE 5.2

Find the steady-state temperature of the thin rod from our analysis of the heat equation if the fixed temperatures at the ends $x = 0$ and $x = \pi$ are w_1 and w_2 respectively.

SOLUTION

The phrase "steady state" means that $\partial u / \partial t = 0$, so that the heat equation reduced to $\partial^2 u / \partial x^2 = 0$ or $d^2 u / dx^2 = 0$. The general solution is then $u = Ax + B$. The values of these two constants are forced by the two boundary conditions; a little high-school algebra tells us that

$$u = w_1 + \frac{1}{\pi}(w_2 - w_1)x.$$

The steady-state version of the three-dimensional heat equation

$$a^2 \left(\frac{\partial^2 u}{\partial x^2} + \frac{\partial^2 u}{\partial y^2} + \frac{\partial^2 u}{\partial z^2} \right) = \frac{\partial u}{\partial t}$$

is

$$\frac{\partial^2 u}{\partial x^2} + \frac{\partial^2 u}{\partial y^2} + \frac{\partial^2 u}{\partial z^2} = 0.$$

This last is called *Laplace's equation*. The study of this equation and its solutions and subsolutions and their applications is a deep and rich branch of mathematics called *potential theory*. There are applications to heat, to gravitation, to electro-magnetics, and to many other parts of physics. The equation plays a central role in the theory of partial differential equations, and is also an integral part of complex variable theory.

Math Note: We now have a good understanding of heat flow in a rod. It is natural to wonder about heat flow in a two-dimensional conductor, such as a disc. Two-dimensional (and higher-dimensional) analysis is quite different from the analysis in one dimension. We shall get a taste of the higher-dimensional tools in the next section.

5.4 The Dirichlet Problem for a Disc

We now study the two-dimensional Laplace equation, which is

$$\triangle u = \frac{\partial^2 u}{\partial x^2} + \frac{\partial^2 u}{\partial y^2} = 0.$$

It will be useful for us to write this equation in polar coordinates. To do so, recall that

$$r^2 = x^2 + y^2, \quad x = r \cos\theta, \quad y = r \sin\theta.$$

Thus

$$\frac{\partial}{\partial r} = \frac{\partial x}{\partial r} \frac{\partial}{\partial x} + \frac{\partial y}{\partial r} \frac{\partial}{\partial y} = \cos\theta \frac{\partial}{\partial x} + \sin\theta \frac{\partial}{\partial y},$$

$$\frac{\partial}{\partial \theta} = \frac{\partial x}{\partial \theta} \frac{\partial}{\partial x} + \frac{\partial y}{\partial \theta} \frac{\partial}{\partial y} = -r \sin\theta \frac{\partial}{\partial x} + r \cos\theta \frac{\partial}{\partial y}.$$

We may solve these two equations for the unknowns $\partial/\partial x$ and $\partial/\partial y$. The result is

$$\frac{\partial}{\partial x} = \cos\theta \frac{\partial}{\partial r} - \frac{\sin\theta}{r}\frac{\partial}{\partial\theta} \quad \text{and} \quad \frac{\partial}{\partial y} = \sin\theta \frac{\partial}{\partial r} - \frac{\cos\theta}{r}\frac{\partial}{\partial\theta}.$$

A tedious calculation now reveals that

$$\Delta = \frac{\partial^2}{\partial x^2} + \frac{\partial^2}{\partial y^2}$$

$$= \left(\cos\theta\frac{\partial}{\partial r} - \frac{\sin\theta}{r}\frac{\partial}{\partial\theta}\right)\left(\cos\theta\frac{\partial}{\partial r} - \frac{\sin\theta}{r}\frac{\partial}{\partial\theta}\right)$$

$$+ \left(\sin\theta\frac{\partial}{\partial r} - \frac{\cos\theta}{r}\frac{\partial}{\partial\theta}\right)\left(\sin\theta\frac{\partial}{\partial r} - \frac{\cos\theta}{r}\frac{\partial}{\partial\theta}\right)$$

$$= \frac{\partial^2}{\partial r^2} + \frac{1}{r}\frac{\partial}{\partial r} + \frac{1}{r^2}\frac{\partial^2}{\partial\theta^2}.$$

Let us fall back once again on the separation of variables method. We will seek a solution $w = w(r, \theta) = u(r) \cdot v(\theta)$ of the Laplace equation. Using the polar form, we find that this leads to the equation

$$u''(r) \cdot v(\theta) + \frac{1}{r}u'(r) \cdot v(\theta) + \frac{1}{r^2}u(r) \cdot v''(\theta) = 0.$$

Thus

$$\frac{r^2 u''(r) + r u'(r)}{u(r)} = -\frac{v''(\theta)}{v(\theta)}.$$

Since the left-hand side depends only on r, and the right-hand side only on θ, both sides must be constant. Denote the common constant value by λ.

Then we have

$$v'' + \lambda v = 0 \tag{1}$$

and

$$r^2 u'' + r u' - \lambda u = 0. \tag{2}$$

If we demand that v be continuous and periodic, then we must demand that $\lambda > 0$ and in fact that $\lambda = n^2$ for some nonnegative integer n. We have studied this situation in detail in Section 5.2. For $n = 0$ the only suitable solution is $v \equiv \text{constant}$ and for $n > 0$ the general solution (with $\lambda = n^2$) is

$$y = A\cos n\theta + B\sin n\theta.$$

We set $\lambda = n^2$ in equation (2), and obtain

$$r^2 u'' + r u' - n^2 u = 0,$$

which is Euler's equidimensional equation. The change of variables $x = e^z$ transforms this equation to a linear equation with constant coefficients, and that can in turn be solved with our standard techniques. The result is

$$u = A + B \ln r \qquad \text{if } n = 0;$$

$$u = A r^n + B r^{-n} \qquad \text{if } n = 1, 2, 3, \ldots.$$

We are most interested in solutions u that are continuous at the origin, so we take $B = 0$ in all cases. The resulting solutions are

$$n = 0, \qquad w = \text{a constant } a_0/2;$$

$$n = 1, \qquad w = r(a_1 \cos\theta + b_1 \sin\theta);$$

$$n = 2, \qquad w = r^2(a_2 \cos 2\theta + b_2 \sin 2\theta);$$

$$n = 3, \qquad w = r^3(a_3 \cos 3\theta + b_3 \sin 3\theta);$$

$$\cdots$$

Of course any finite sum of solutions of Laplace's equation is also a solution. The same is true for infinite sums. Thus we are led to consider

$$w = w(r, \theta) = \tfrac{1}{2} a_0 + \sum_{j=1}^{\infty} r^j (a_j \cos j\theta + b_j \sin j\theta).$$

On a formal level, letting $r \to 1^-$ in this last expression gives

$$w = \tfrac{1}{2} a_0 + \sum_{j=1}^{\infty} (a_j \cos j\theta + b_j \sin j\theta).$$

Math Note: We draw all these ideas together with the following physical rubric. Consider a thin aluminum disc of radius 1, and imagine applying a heat distribution to the boundary of that disc. In polar coordinates, this distribution is specified by a function $f(\theta)$. We seek to understand the steady-state heat distribution on the entire disc. So we seek a function $w(r, \theta)$, continuous on the closure of the disc, which agrees with f on the boundary and which represents the steady-state distribution of heat inside. Some physical analysis shows that such a function w is the solution of the boundary value problem

$$\triangle w = 0,$$

$$w\bigg|_{\partial D} = f.$$

According to the calculations we performed above, a natural approach to this problem is to expand the given function f in its Fourier series:

$$f(\theta) = \tfrac{1}{2}a_0 + \sum_{j=1}^{\infty}(a_j \cos j\theta + b_j \sin j\theta)$$

and then posit that the w we seek is

$$w(r, \theta) = \tfrac{1}{2}a_0 + \sum_{j=1}^{\infty}r^j(a_j \cos j\theta + b_j \sin j\theta).$$

This process is known as solving *the Dirichlet problem on the disc with boundary data f.*

EXAMPLE 5.3
Follow the paradigm just sketched to solve the Dirichlet problem on the disc with $f(\theta) = 1$ on the top half of the boundary and $f(\theta) = -1$ on the bottom half of the boundary.

SOLUTION
It is straightforward to calculate that the Fourier series (sine series) expansion for this f is

$$f(\theta) = \frac{4}{\pi}\left(\sin\theta + \frac{\sin 3\theta}{3} + +\frac{\sin 5\theta}{5} + \cdots\right).$$

The solution of the Dirichlet problem is therefore

$$w(r, \theta) = \frac{4}{\pi}\left(r\sin\theta + \frac{r^3\sin 3\theta}{3} + +\frac{r^5\sin 5\theta}{5} + \cdots\right).$$

You Try It: Solve the Dirichlet problem on the disc with boundary data $f(\theta) = \theta$, $0 \le \theta < 2\pi$.

5.4.1 THE POISSON INTEGRAL

We have presented a formal procedure with series for solving the Dirichlet problem. But in fact it is possible to produce a closed formula for this solution. This we now do.

Referring back to our sine series expansion for f, and the resulting expansion for the solution of the Dirichlet problem, we recall that

$$a_j = \frac{1}{\pi} \int_{-\pi}^{\pi} f(\phi) \cos j\phi \, d\phi \quad \text{and} \quad b_j = \frac{1}{\pi} \int_{-\pi}^{\pi} f(\phi) \sin j\phi \, d\phi.$$

Thus

$$w(r, \theta) = \frac{1}{2} a_0 + \sum_{j=1}^{\infty} r^j \left(\frac{1}{\pi} \int_{-\pi}^{\pi} f(\phi) \cos j\phi \, d\phi \right) \cos j\theta$$

$$+ \left(\frac{1}{\pi} \int_{-\pi}^{\pi} f(\phi) \sin j\phi \, d\phi \sin j\theta \right).$$

This, in turn, equals

$$\frac{1}{2} a_0 + \frac{1}{\pi} \sum_{j=1}^{\infty} r^j \int_{-\pi}^{\pi} f(\phi) \big[\cos j\phi \cos j\theta$$

$$+ \sin j\phi \sin j\theta d\phi \big]$$

$$= \frac{1}{2} a_0 + \frac{1}{\pi} \sum_{j=1}^{\infty} r^j \int_{-\pi}^{\pi} f(\phi) \left[\cos j(\theta - \phi) d\phi \right].$$

We finally simplify our expression to

$$w(r, \theta) = \frac{1}{\pi} \int_{-\pi}^{\pi} f(\phi) \left[\frac{1}{2} + \sum_{j=1}^{\infty} r^j \cos j(\theta - \phi) \right] d\phi.$$

It behooves us, therefore, to calculate the sum inside the brackets. For simplicity, we let $\alpha = \theta - \phi$ and then we let

$$z = re^{i\alpha} = r(\cos \alpha + i \sin \alpha).$$

Likewise

$$z^n = r^n e^{in\alpha} = r^n (\cos n\alpha + i \sin n\alpha).$$

Let Re z denote the real part of the complex number z. Then

$$\frac{1}{2} + \sum_{j=1}^{\infty} r^j \cos j\alpha = \text{Re}\left[\frac{1}{2} + \sum_{j=1}^{\infty} z^j\right]$$

$$= \text{Re}\left[-\frac{1}{2} + \frac{1}{1-z}\right]$$

$$= \text{Re}\left[\frac{1+z}{2(1-z)}\right]$$

$$= \text{Re}\left[\frac{(1+z)(1-\bar{z})}{2|1-z|^2}\right]$$

$$= \frac{1-|z|^2}{2|1-z|^2}$$

$$= \frac{1-r^2}{2(1-2r\cos\alpha + r^2)}.$$

Putting the result of this calculation into our original formula for w we finally obtain the Poisson integral formula:

$$w(r,\theta) = \frac{1}{2\pi}\int_{-\pi}^{\pi} \frac{1-r^2}{1-2r\cos\alpha + r^2} f(\phi)\,d\phi.$$

Observe what this formula does for us: It expresses the solution of the Dirichlet problem with boundary data f as an explicit integral of a universal expression (called a *kernel*) against that data function f.

There is a great deal of information about w and its relation to f contained in this formula. As just one simple instance, we note that when r is set equal to 0, we obtain

$$w(0,\theta) = \frac{1}{2\pi}\int_{-\pi}^{\pi} f(\phi)\,d\phi.$$

This says that the value of the steady-state heat distribution at the origin is just the average value of f around the circular boundary.

Math Note: The Poisson kernel (and integral) is but one example of a *reproducing kernel* in mathematics. There are many others—the Cauchy kernel, the Bergman kernel, and the Szegö among them. These are powerful tools for analyzing and continuing (or extending) functions.

5.5 Sturm–Liouville Problems

We wish to place the idea of eigenvalues and eigenfunctions into a broader context. This setting is the fairly broad and far-reaching subject of Sturm–Liouville problems.

Recall that a sequence y_j of functions such that

$$\int_a^b y_m(x) y_n(x)\, dx = 0 \qquad \text{for } m \neq n$$

is said to be an *orthogonal system* on the interval $[a, b]$. If

$$\int_a^b y_j^2(x)\, dx = 1$$

for each j, then we call our collection of functions an *orthonormal system* or *orthonormal sequence*. It turns out (and we have seen several instances of this phenomenon) that the sequence of eigenfunctions associated with a wide variety of boundary value problems enjoys the orthogonality property.

Now consider a differential equation of the form

$$\frac{d}{dx}\left[p(x)\frac{dy}{dx}\right] + [\lambda q(x) + r(x)]y = 0; \tag{1}$$

we shall be interested in solutions valid on an interval $[a, b]$. We know that, under suitable conditions on the coefficients, a solution of this equation (1) that takes a prescribed value and a prescribed derivative value at a fixed point $x_0 \in [a, b]$ will be uniquely determined. In other circumstances, we may wish to prescribe the values of y at two distinct points, say at a and at b. We now begin to examine the conditions under which such a *boundary value problem* has a nontrivial solution.

 EXAMPLE 5.4

Consider the equation (1) with $p(x) \equiv q(x) \equiv 1$ and $r(x) \equiv 0$. Then the differential equation becomes

$$y'' + \lambda y = 0.$$

We take the domain interval to be $[0, \pi]$ and the boundary conditions to be

$$y(0) = 0, \quad y(\pi) = 0.$$

What are the eigenvalues and eigenfunctions for this problem?

SOLUTION
Of course we completely analyzed this problem in Section 5.2. But now, as motivation for the work in this section, we review. We know that, in order for this boundary value problem to have a solution, the parameter λ can only assume the values $\lambda_n = n^2$, $n = 1, 2, 3, \ldots$. The corresponding solutions to the differential equation are $y_n(x) = \sin nx$. We call λ_n the *eigenvalues* for the problem and y_n the *eigenfunctions* (or sometimes the *eigenvectors*) for the problem.

You Try It: Consider the differential equation

$$y'' + \lambda y = 0.$$

We take the domain interval to be $[0, \pi]$ and the boundary conditions to be

$$y(0) = 0, \quad y(\pi) = 0.$$

What are the eigenvalues and eigenfunctions for this problem?

It will turn out—and this is the basis for the Sturm–Liouville theory—that if $p, q > 0$ on $[a, b]$, then the equation (1) will have a solvable boundary value problem—for a certain discrete set of values of λ—with data specified at points a and b. These special values of λ will of course be the eigenvalues for the boundary value problem. They are real numbers that we shall arrange in their natural order

$$\lambda_1 < \lambda_2 < \cdots < \lambda_n < \cdots,$$

and we shall learn that $\lambda_j \to +\infty$. The corresponding eigenfunctions will then be ordered as $y_1, y_2, \ldots$.

Now let us examine possible orthogonality properties for the eigenfunctions of the boundary value problem for equation (1). Consider the differential equation (1) with two different eigenvalues λ_m and λ_n and y_m and y_n the corresponding eigenfunctions:

$$\frac{d}{dx}\left[p(x)\frac{dy_m}{dx}\right] + [\lambda q(x) + r(x)]y_m = 0$$

and

$$\frac{d}{dx}\left[p(x)\frac{dy_n}{dx}\right] + [\lambda q(x) + r(x)]y_n = 0.$$

We convert to the more convenient prime notation for derivatives, multiply the first equation by y_n and the second by y_m, and subtract. The result is

$$y_n(py_m')' - y_m(py_n')' + (\lambda_m - \lambda_n)qy_my_n = 0.$$

We move the first two terms to the right-hand side of the equation and integrate from a to b. Hence

$$(\lambda_m - \lambda_n) \int_a^b q\, y_m\, y_n\, dx$$

$$= \int_a^b y_m (p y_n')'\, dx - \int_a^b y_n (p y_m')'\, dx$$

$$\overset{\text{(parts)}}{=} \left[y_m (p y_n') \right]_a^b - \int_a^b y_m' (p y_n')\, dx$$

$$- \left[y_n (p y_m') \right]_a^b + \int_a^b y_n' (p y_m')\, dx$$

$$= p(b)[y_m(b) y_n'(b) - y_n(b) y_m'(b)]$$

$$- p(a)[y_m(a) y_n'(a) - y_n(a) y_m'(a)].$$

Let us denote by $W(x)$ the Wronskian determinant[1] of the two solutions y_m, y_n. Thus

$$W(x) = y_m(x) y_n'(x) - y_n(x) y_m'(x).$$

Then our last equation can be written in the more compact form

$$(\lambda_m - \lambda_n) \int_a^b q\, y_m\, y_n\, dx = p(b) W(b) - p(a) W(a).$$

Notice that things have turned out so nicely, and certain terms have cancelled, just because of the special form of the original differential equation.

We want the right-hand side of this last equation to vanish. This will certainly be the case if we require the familiar boundary conditions

$$y(a) = 0 \qquad \text{and} \qquad y(b) = 0$$

or instead we require that

$$y'(a) = 0 \qquad \text{and} \qquad y'(b) = 0.$$

Either of these will guarantee that the Wronskian vanishes, and therefore

$$\int_a^b y_m \cdot y_n \cdot q\, dx = 0.$$

This is called an *orthogonality condition with weight q*.

---

[1] It is a fact that the Wronskian is either identically 0 or never 0. In the second instance, we may conclude that y_m, y_n are linearly independent. Otherwise they are linearly dependent.

With such a condition in place, we can consider representing an arbitrary function f as a linear combination of the y_j:

$$f(x) = a_1 y_1(x) + a_2 y_2(x) + \cdots + a_j y_j(x) + \cdots . \qquad (2)$$

We may determine the coefficients a_j by multiplying both sides of this equation by $y_k \cdot q$ and integrating from a to b. Thus

$$\int_a^b f(x) y_k(x) q(x)\, dx = \int_a^b \big[a_1 y_1(x) + a_2 y_2(x) + \cdots$$

$$+ a_j y_j(x) + \cdots \big] y_k(x) q(x)\, dx$$

$$= \sum_j a_j \int_a^b y_j(x) y_k(x) q(x)\, dx$$

$$= a_k \int_a^b y_k^2(x) q(x)\, dx.$$

Thus

$$a_k = \frac{\int_a^b f(x) y_k(x) q(x)\, dx}{\int_a^b y_k^2(x) q(x)\, dx}.$$

Math Note: You should notice the parallel between these calculations and the ones we performed in Subsection 5.2.3. The idea of orthogonality with respect to a weight has now arisen for us in a concrete context. Certainly Sturm–Liouville problems play a prominent role in engineering problems, especially ones coming from mechanics.

There is an important question that now must be asked. Namely, are there *enough* of the eigenfunctions y_j so that virtually any function f can be expanded as in (2)? For instance, the functions $y_1(x) = \sin x$, $y_3(x) = \sin 3x$, $y_7(x) = \sin 7x$ are orthogonal on $[-\pi, \pi]$, and for any function f one can calculate coefficients a_1, a_3, a_7. But there is no hope that a large class of functions f can be spanned by just y_1, y_3, y_7. We need to know that our y_j's "fill out the space." The study of this question is beyond the scope of the present text, as it involves ideas from Hilbert space (see [RUD]). Our intention here has been merely to acquaint the reader with some of the language of Sturm–Liouville problems.

Exercises

1. Find the eigenvalues λ_n and the eigenfunctions y_n for the equation $y'' + \lambda y = 0$ in each of the following instances:
   (a) $y(0) = 0,$ $y(\pi/2) = 0$
   (b) $y(0) = 0,$ $y(2\pi) = 0$
   (c) $y(0) = 0,$ $y(1) = 0$

2. Solve the vibrating string problem in the text if the initial shape $y(x, 0) = f(x)$ is specified by the given function. In each case, sketch the initial shape of the string on a set of axes.

   (a) $f(x) = \begin{cases} 2x/\pi & \text{if } 0 \le x \le \pi/2 \\ 2(\pi - x)/\pi & \text{if } \pi/2 < x \le \pi \end{cases}$

   (b) $f(x) = \dfrac{1}{\pi} x(\pi - x)$

3. Solve the vibrating string problem in the text if the initial shape $y(x, 0) = f(x)$ is that of a single arch of the sine curve $f(x) = c \sin x$. Show that the moving string has the same general shape, regardless of the value of c.

4. The problem of the *struck string* is that of solving the wave equation with the boundary conditions

$$y(0, t) = 0, \qquad y(\pi, t) = 0$$

   and the initial conditions

$$\left. \frac{\partial y}{\partial t} \right|_{t=0} = g(x) \qquad \text{and} \qquad y(x, 0) = 0.$$

   [These initial conditions mean that the string is initially in the equilibrium position, and has an initial velocity $g(x)$ at the point x as a result of being struck.] By separating variables and proceeding formally, obtain the solution

$$y(x, t) = \sum_{j=1}^{\infty} c_j \sin jx \sin jat.$$

5. Solve the problem of finding $w(x, t)$ for the rod with insulated ends at $x = 0$ and $x = \pi$ (with temperatures held at 0 degrees) if the initial temperature distribution is given by $w(x, 0) = f(x)$.

6. Solve the Dirichlet problem for the unit disc when the boundary function $f(\theta)$ is defined by
    (a) $f(\theta) = \cos\theta/2, \quad -\pi \le \theta \le \pi$
    (b) $f(\theta) = \theta, \quad -\pi \le \theta \le \pi$
    (c) $f(\theta) = \begin{cases} 0 & \text{if } -\pi \le \theta < 0 \\ \sin\theta & \text{if } 0 \le \theta \le \pi \end{cases}$

7. Show that the Dirichlet problem for the disc $\{(x, y) : x^2 + y^2 \le R^2\}$, where $f(\theta)$ is the boundary function, has the solution

    $$w(r, \theta) = \frac{1}{2}a_0 + \sum_{j=1}^{\infty} \left(\frac{r}{R}\right)^j (a_j \cos j\theta + b_j \sin j\theta)$$

    where a_j and b_j are the Fourier coefficients of f.

8. Solve the vibrating string problem if the initial shape $y(x, 0) = f(x)$ is specified by the function

    $$f(x) = \begin{cases} x & \text{if } 0 \le x \le \dfrac{\pi}{2} \\ \pi - x & \text{if } \dfrac{\pi}{2} < x \le \pi. \end{cases}$$

9. Solve the Dirichlet problem for the unit disc when the boundary function $f(\theta)$ is defined by $f(\theta) = \theta - |\theta|, \quad -\pi \le \theta \le \pi$.

CHAPTER 6

Laplace Transforms

6.1 Introduction

The idea of the Laplace transform has had a profound influence over the development of mathematical analysis. It also plays a significant role in mathematical applications. More generally, the overall theory of transforms has become an important part of modern mathematics.

The idea of a *transform* is that it turns a given function into another function. We are already acquainted with several transforms:

I.   The derivative D takes a differentiable function f (defined on some interval (a, b)) and assigns to it a new function $Df = f'$.

II.   The integral I takes a continuous function f (defined on some interval $[a, b]$ and assigns to it a new function

$$If(x) = \int_a^x f(t)\,dt.$$

III.   The multiplication operator M_φ, which multiplies any given function f on the interval $[a, b]$ by a fixed function φ on $[a, b]$, is a transform:

$$M_\varphi f(x) = \varphi(x) \cdot f(x).$$

We are particularly interested in transforms that are linear. A transform T is *linear* if

$$T[\alpha f + \beta g] = \alpha T(f) + \beta T(g)$$

for any real constants α, β. In particular (taking $\alpha = \beta = 1$),

$$T[f + g] = T(f) + T(g)$$

and (taking $\beta = 0$)

$$T(\alpha f) = \alpha T(f).$$

We can most fruitfully study linear transformations that are given by integration. The *Laplace transform* is defined by

$$L[f](p) = \int_0^\infty e^{-px} f(x) \, dx \qquad \text{for } p > 0.$$

Notice that we begin with a function f of x, and the Laplace transform L produces a new function $L[f]$ of p. We sometimes write the Laplace transform of $f(x)$ as $F(p)$. Notice that the Laplace transform is an improper integral; it exists precisely when

$$\int_0^\infty e^{-px} f(x) \, dx = \lim_{N \to \infty} \int_0^N e^{-px} f(x) \, dx$$

exists.

Let us now calculate some Laplace transforms:

EXAMPLE 6.1
Calculate the Laplace transform of x^n.

SOLUTION

$$L[x^n] = \int_0^\infty e^{-px} x^n \, dx$$

$$\overset{\text{(parts)}}{=} -\frac{x^n e^{-px}}{p}\Big|_0^\infty + \frac{n}{p} \int_0^\infty e^{-px} x^{n-1} \, dx$$

$$= \frac{n}{p} L[x^{n-1}]$$

$$= \frac{n}{p} \left(\frac{n-1}{p} \right) L[x^{n-2}]$$

Table 6.1

Function f	Laplace transform F
$f(x) \equiv 1$	$F(p) = \int_0^\infty e^{-px}\, dx = \frac{1}{p}$
$f(x) = x$	$F(p) = \int_0^\infty e^{-px} x\, dx = \frac{1}{p^2}$
$f(x) = x^n$	$F(p) = \int_0^\infty e^{-px} x^n\, dx = \frac{n!}{p^{n+1}}$
$f(x) = e^{ax}$	$F(p) = \int_0^\infty e^{-px} e^{ax}\, dx = \frac{1}{p-a}$
$f(x) = \sin ax$	$F(p) = \int_0^\infty e^{-px} \sin ax\, dx = \frac{a}{p^2+a^2}$
$f(x) = \cos ax$	$F(p) = \int_0^\infty e^{-px} \cos ax\, dx = \frac{p}{p^2+a^2}$
$f(x) = \sinh ax$	$F(p) = \int_0^\infty e^{-px} \sinh ax\, dx = \frac{a}{p^2-a^2}$
$f(x) = \cosh ax$	$F(p) = \int_0^\infty e^{-px} \cosh ax\, dx = \frac{p}{p^2-a^2}$

$$= \cdots = \frac{n!}{p^n} L[1]$$

$$= \frac{n!}{p^{n+1}}.$$

You will find, as we have just seen, that integration by parts is eminently useful in the calculation of Laplace transforms.

We shall not actually perform all the integrations for the Laplace transforms in Table 6.1. We content ourselves with the third one, just to illustrate the idea. You should definitely perform the others, just to get the feel of Laplace transform calculations.

☞ **You Try It:** Calculate the Laplace transform of $\sin ax$.

It may be noted that the Laplace transform is a linear operator. Thus Laplace transforms of some compound functions may be readily calculated from the table just given:

$$L(5x^3 - 2e^x) = \frac{5 \cdot 3!}{p^4} - \frac{2}{p - 1}$$

and

$$L(4\sin 2x + 6x) = \frac{4 \cdot 2}{p^2 + 2^2} + \frac{6}{p^2}.$$

6.2 Applications to Differential Equations

The key to our use of Laplace transform theory in the subject of differential equations is the way that L treats derivatives. Let us calculate

$$L[y'] = \int_0^\infty e^{-px} y'(x)\, dx$$

$$\overset{\text{(parts)}}{=} -ye^{-px}\Big|_0^\infty + p \int_0^\infty e^{-px} y\, dx$$

$$= -y(0) + p \cdot L[y].$$

In summary,

$$L[y'] = p \cdot L[y] - y(0). \tag{1}$$

Likewise,

$$L[y''] = L[(y')'] = p \cdot L[y'] - y'(0)$$

$$= p\{p \cdot L[y] - y(0)\} - y'(0)$$

$$= p^2 \cdot L[y] - py(0) - y'(0). \tag{2}$$

Now let us examine the differential equation

$$y'' + ay' + by = f(x), \tag{3}$$

with the initial conditions $y(0) = y_0$ and $y'(0) = y_1$. Here a and b are real constants. We apply the Laplace transform L to both sides of (3), of course using the linearity of L. The result is

$$L[y''] + aL[y'] + bL[y] = L[f].$$

Writing out what each term is, we find that

$$p^2 \cdot L[y] - py(0) - y'(0) + a\{p \cdot L[y] - y(0)\} + bL[y] = L[f].$$

Now we can plug in what $y(0)$ and $y'(0)$ are. We may also gather like terms together. The result is

$$\{p^2 + ap + b\}L[y] = (p + a)y_0 + y_1 + L[f]$$

or

$$L[y] = \frac{(p + a)y_0 + y_1 + L[f]}{p^2 + ap + b}. \tag{4}$$

What we see here is a remarkable thing: The Laplace transform changes solving a differential equation from a rather complicated calculus problem to a simple algebra problem. The only thing that remains, in order to find an explicit solution to the original differential equation (3), is to find the inverse Laplace transform of the right-hand side of (4). In practice we will find that we can often perform this operation in a straightforward fashion. The following examples will illustrate the idea.

EXAMPLE 6.2

Use the Laplace transform to solve the differential equation

$$y'' + 4y = 4x \tag{5}$$

with initial conditions $y(0) = 1$ and $y'(0) = 5$.

SOLUTION

We proceed mechanically, by applying the Laplace transform to both sides of (5). Thus

$$L[y''] + L[4y] = L[4x].$$

We can use our various Laplace transform formulas to write this out more explicitly:

$$\{p^2 L[y] - py(0) - y'(0)\} + 4L[y] = \frac{4}{p^2}$$

or

$$p^2 L[y] - p \cdot 1 - 5 + 4L[y] = \frac{4}{p^2}$$

or

$$(p^2 + 4)L[y] = p + 5 + \frac{4}{p^2}.$$

It is convenient to write this as

$$L[y] = \frac{p}{p^2 + 4} + \frac{5}{p^2 + 4} + \frac{4}{p^2 \cdot (p^2 + 4)}$$

$$= \frac{p}{p^2 + 4} + \frac{5}{p^2 + 4} + \left[\frac{1}{p^2} - \frac{1}{p^2 + 4} \right],$$

where we have used a partial fractions decomposition in the last step. Simplifying, we have

$$L[y] = \frac{p}{p^2 + 4} + \frac{4}{p^2 + 4} + \frac{1}{p^2}.$$

Referring to our table of Laplace transforms, we may now deduce what y must be:

$$L[y] = L[\cos 2x] + L[2 \sin 2x] + L[x] = L[\cos 2x + 2 \sin 2x + x].$$

Now it is known that the Laplace transform is one-to-one: if $L[f] = L[g]$, then $f = g$. Using this important property, we deduce then that

$$y = \cos 2x + 2 \sin 2x + x,$$

and this is the solution of our initial value problem.

A useful general property of the Laplace transform concerns its interaction with translations. Indeed, we have

$$L[e^{ax} f(x)] = F(p - a). \tag{6}$$

To see this, we calculate

$$L[e^{ax} f(x)] = \int_0^\infty e^{-px} e^{ax} f(x) \, dx$$

$$= \int_0^\infty e^{-(p-a)x} f(x) \, dx$$

$$= F(p - a).$$

We frequently find it useful to use the notation L^{-1} to denote the inverse operation to the Laplace transform.[1] For example, since

$$L[x^2] = \frac{2!}{p^3},$$

we may write

$$L^{-1}\left[\frac{2!}{p^3}\right] = x^2.$$

---

[1] We tacitly use here the fact that the Laplace transform L is one-to-one: if $L[f] = L[g]$, then $f = g$. Thus L is invertible on its image. We are able to verify this assertion empirically through our calculations; the general result is proved in a more advanced treatment.

Since

$$L[\sin x - e^{2x}] = \frac{1}{p^2 + 1} - \frac{1}{p - 2},$$

we may write

$$L^{-1}\left[\frac{1}{p^2 + 1} - \frac{1}{p - 2}\right] = \sin x - e^{2x}.$$

EXAMPLE 6.3
Since

$$L[\sin bx] = \frac{b}{p^2 + b^2},$$

we conclude that

$$L[e^{ax}\sin bx] = \frac{b}{(p - a)^2 + b^2}.$$

Since

$$L^{-1}\left[\frac{1}{p^2}\right] = x,$$

we conclude that

$$L^{-1}\left[\frac{1}{(p - a)^2}\right] = e^{ax}x.$$

EXAMPLE 6.4
Use the Laplace transform to solve the differential equation

$$y'' + 2y' + 5y = 3e^{-x}\sin x \tag{7}$$

with initial conditions $y(0) = 0$ and $y'(0) = 3$.

SOLUTION
We calculate the Laplace transform of both sides, using our new formula (6) on the right-hand side, to obtain

$$\left\{p^2 L[y] - py(0) - y'(0)\right\} + 2\left\{pL[y] - y(0)\right\} + 5L[y] = 3 \cdot \frac{1}{(p + 1)^2 + 1}.$$

Plugging in the initial conditions, and organizing like terms, we find that

$$(p^2 + 2p + 5)L[y] = 3 + \frac{3}{(p + 1)^2 + 1}$$

or

$$L[y] = \frac{3}{p^2 + 2p + 5} + \frac{3}{(p^2 + 2p + 2)(p^2 + 2p + 5)}$$

$$= \frac{3}{p^2 + 2p + 5} + \frac{1}{p^2 + 2p + 2} - \frac{1}{p^2 + 2p + 5}$$

$$= \frac{2}{(p + 1)^2 + 4} + \frac{1}{(p + 1)^2 + 1}.$$

We see therefore that

$$y = e^{-x} \sin 2x + e^{-x} \sin x.$$

This is the solution of our initial value problem.

You Try It:　Use the Laplace transform to solve the differential equation

$$y'' + y' + y = e^x.$$

Math Note:　Since we know how to calculate the Laplace transform of the derivative of a function, it is natural also to consider the Laplace transform for the *anti*derivative of a function. Derive a suitable formula.

6.3　Derivatives and Integrals of Laplace Transforms

In some contexts it is useful to calculate the derivative of the Laplace transform of a function (when the corresponding integral make sense). For instance, consider

$$F(p) = \int_0^\infty e^{-px} f(x)\, dx.$$

Then

$$\frac{d}{dp} F(p) = \frac{d}{dp} \int_0^\infty e^{-px} f(x)\, dx$$

$$= \int_0^\infty \frac{d}{dp} e^{-px} f(x)\, dx$$

$$= \int_0^\infty e^{-px} \{-x f(x)\}\, dx = L[-x f(x)](p).$$

We see that the derivative[2] of $F(p)$ is the Laplace transform of $-xf(x)$. More generally, the same calculation shows us that

$$\frac{d^2}{dp^2}F(p) = L[x^2 f(x)](p)$$

and

$$\frac{d^j}{dp^j}F(p) = L[(-1)^j x^j f(x)](p).$$

 EXAMPLE 6.5
Calculate

$$L[x \sin ax].$$

SOLUTION
We have

$$L[x \sin ax] = -L[-x \sin ax] = -\frac{d}{dp}L[\sin ax] = -\frac{d}{dp}\left(\frac{a}{p^2+a^2}\right) = \frac{2ap}{(p^2+a^2)^2}.$$

 EXAMPLE 6.6
Calculate the Laplace transform of $\sqrt{x}$.

SOLUTION
This calculation actually involves some tricky integration. We first note that

$$L[\sqrt{x}] = L[x^{1/2}] = -L[-x \cdot x^{-1/2}] = -\frac{d}{dp}L[x^{-1/2}]. \qquad (1)$$

Thus we must find the Laplace transform of $x^{-1/2}$.
    Now

$$L[x^{-1/2}] = \int_0^\infty e^{-px}x^{-1/2}\,dx.$$

The change of variables $px = t$ yields

$$= p^{-1/2}\int_0^\infty e^{-t}t^{-1/2}\,dt.$$

---

[2]The passage of the derivative under the integral sign in this calculation requires advanced ideas from real analysis that we cannot treat here—see [KRA2].

The further change of variables $t = s^2$ gives the integral

$$L[x^{-1/2}] = 2p^{-1/2} \int_0^\infty e^{-s^2} \, ds. \tag{2}$$

Now we must evaluate the integral $I = \int_0^\infty e^{-s^2} \, ds$. Observe that

$$I \cdot I = \int_0^\infty e^{-s^2} \, ds \cdot \int_0^\infty e^{-u^2} \, du = \int_0^\infty \int_0^{\pi/2} e^{-r^2} \cdot r \, d\theta dr.$$

Here we have introduced polar coordinates in the standard way.
Now the last integral is easily evaluated and we find that

$$I^2 = \frac{\pi}{4},$$

hence $I = \sqrt{\pi}/2$. Thus $L[x^{-1/2}](p) = 2p^{-1/2}\{\sqrt{\pi}/2\} = \sqrt{\pi/p}$. Finally,

$$L[\sqrt{x}] = -\frac{d}{dp}\sqrt{\frac{\pi}{p}} = \frac{1}{2p}\sqrt{\frac{\pi}{p}}.$$

We now derive some additional formulas that will be useful in solving differential equations. We let $y = f(x)$ be our function and $Y = L[f]$ be its Laplace transform. Then

$$L[xy] = -\frac{d}{dp}L[y] = -\frac{dY}{dp}. \tag{3}$$

Also

$$L[xy'] = -\frac{d}{dp}L[y'] = -\frac{d}{dp}[pY - y(0)] = -\frac{d}{dp}[pY] \tag{4}$$

and

$$L[xy''] = -\frac{d}{dp}L[y''] = -\frac{d}{dp}[p^2 Y - py(0) - y'(0)] = -\frac{d}{dp}[p^2 Y - py(0)]. \tag{5}$$

EXAMPLE 6.7

Use the Laplace transform to analyze Bessel's equation

$$xy'' + y' + xy = 0$$

with the single initial condition $y(0) = 1$.

SOLUTION

Apply the Laplace transform to both sides of the equation. Thus

$$L[xy''] + L[y'] + L[xy] = L[0] = 0.$$

We can apply our new formulas (5) and (3) to the first and third terms on the left. And of course we apply the usual form for the Laplace transform of the derivative to the second term on the left. The result is

$$-\frac{d}{dp}[p^2Y - p] + \{pY - 1\} + \left\{-1 - \frac{dY}{dp}\right\} = 0.$$

We may simplify this equation to

$$(p^2 + 1)\frac{dY}{dp} = -pY.$$

This is a *new* differential equation, and we may solve it by separation of variables. Now

$$\frac{dY}{Y} = -\frac{p\,dp}{p^2 + 1},$$

so

$$\ln Y = -\frac{1}{2}\ln(p^2 + 1) + C.$$

Exponentiating both sides gives

$$Y = D \cdot \frac{1}{\sqrt{p^2 + 1}}.$$

It is convenient (with a view to calculating the inverse Laplace transform) to write this solution as

$$Y = \frac{D}{p} \cdot \left(1 + \frac{1}{p^2}\right)^{-1/2}. \tag{6}$$

Recall the binomial expansion

$$(1 + z)^a = 1 + az + \frac{a(a-1)}{2!} + \frac{a(a-1)(a-2)}{3!}$$

$$+ \cdots + \frac{a(a-1)\cdots(a-n+1)}{n!} + \cdots.$$

We apply this formula to the second term on the right of (6). Thus

$$Y = \frac{D}{p} \cdot \left[1 - \frac{1}{2} \cdot \frac{1}{p^2} + \frac{1}{2!} \cdot \frac{1}{2} \cdot \frac{3}{2} \cdot \frac{1}{p^4} - \frac{1}{3!} \cdot \frac{1}{2} \cdot \frac{3}{2} \cdot \frac{5}{2} \cdot \frac{1}{p^6} \right.$$

$$\left. + \cdots + \frac{1 \cdot 3 \cdot 5 \cdots (2n-1)}{2^n n!} \frac{(-1)^n}{p^{2n}} + \cdots \right]$$

$$= D \cdot \sum_{j=0}^{\infty} \frac{(2j)!}{2^{2j}(j!)^2} \cdot \frac{(-1)^j}{p^{2j+1}}.$$

The good news is that we can now calculate L^{-1} of Y (thus obtaining y) by just calculating the inverse Laplace transform of each term of this series. The result is

$$y(x) = D \cdot \sum_{j=0}^{\infty} \frac{(-1)^j}{2^{2j}(j!)^2} \cdot x^{2j}$$

$$= D \cdot \left(1 - \frac{x^2}{2^2} + \frac{x^4}{2^2 \cdot 4^2} - \frac{x^6}{2^2 \cdot 4^2 \cdot 6^2} + \cdots \right).$$

Since $y(0) = 1$ (the initial condition), we see that $D = 1$ and

$$y(x) = 1 - \frac{x^2}{2^2} + \frac{x^4}{2^2 \cdot 4^2} - \frac{x^6}{2^2 \cdot 4^2 \cdot 6^2} + \cdots .$$

The series we have just derived defines the celebrated and important *Bessel function* J_0. We have learned that the Laplace transform of J_0 is $1/\sqrt{p^2 + 1}$.

You Try It: Use the Laplace transform to solve the differential equation

$$xy'' - xy = \cos x.$$

You Try It: Use the Laplace transform to solve the initial value problem

$$xy'' + xy = 0, \quad y(0) = 1, \quad y'(0) = 0.$$

It is also a matter of some interest to integrate the Laplace transform. We can anticipate how this will go by running the differentiation formulas in reverse. Our main result is

$$L\left[\frac{f(x)}{x} \right] = \int_p^{\infty} F(s) \, ds. \tag{7}$$

In fact

$$\int_p^\infty F(s)\,ds = \int_p^\infty \left[\int_0^\infty e^{-sx} f(x)\,dx\right] ds$$

$$= \int_0^\infty f(x) \int_p^\infty e^{-sx}\,ds\,dx$$

$$= \int_0^\infty f(x) \left[\frac{e^{-sx}}{-x}\right]_p^\infty dx$$

$$= \int_0^\infty f(x) \cdot \frac{e^{-px}}{x}\,dx$$

$$= \int_0^\infty \left[\frac{f(x)}{x}\right] \cdot e^{-px}\,dx$$

$$= L\left[\frac{f(x)}{x}\right].$$

EXAMPLE 6.8

Use the fact that $L[\sin x] = 1/(p^2 + 1)$ to calculate $\int_0^\infty (\sin x)/x\,dx$.

SOLUTION

By formula (7) (with $f(x) = \sin x$),

$$\int_0^\infty \frac{\sin x}{x}\,dx = \int_0^\infty \frac{dp}{p^2 + 1} = \arctan p \Big|_0^\infty = \frac{\pi}{2}.$$

We conclude this section by summarizing the chief properties of the Laplace transform in Table 6.2. The last property listed concerns convolution, and we shall treat that topic in the next section.

6.4 Convolutions

An interesting question, which occurs frequently with the use of the Laplace transform, is this: Let f and g be functions and F and G their Laplace transforms;

Table 6.2   Properties of the Laplace transform

$L[\alpha f(x) + \beta g(x)] = \alpha F(p) + \beta G(p)$
$L[e^{ax} f(x)] = F(p - a)$
$L[f'(x)] = pF(p) - f(0)$
$L[f''(x)] = p^2 F(p) - pf(0) - f'(0)$
$L\left[\int_0^x f(t)\,dt\right] = \dfrac{F(p)}{p}$
$L[-xf(x)] = F'(p)$
$L[(-1)^n x^n f(x)] = F^{(n)}(p)$
$L\left[\dfrac{f(x)}{x}\right] = \int_p^\infty F(p)\,dp$
$L\left[\int_0^x f(x - t)g(t)\,dt\right] = F(p)G(p)$

what is $L^{-1}[F \cdot G]$? To discover the answer, we write

$$F(p) \cdot G(p) = \left[\int_0^\infty e^{-ps} f(s)\,ds\right] \cdot \left[\int_0^\infty e^{-pt} f(t)\,dt\right]$$

$$= \int_0^\infty \int_0^\infty e^{-p(s+t)} f(s)g(t)\,ds\,dt$$

$$= \int_0^\infty \left[\int_0^\infty e^{-p(s+t)} f(s)\,ds\right] g(t)\,dt.$$

Now we perform the change of variable $s = x - t$ in the inner integral. The result is

$$F(p) \cdot G(p) = \int_0^\infty \left[\int_t^\infty e^{-px} f(x - t)\,dx\right] g(t)\,dt$$

$$= \int_0^\infty \int_t^\infty e^{-px} f(x - t)g(t)\,dx\,dt.$$

Reversing the order of integration, we may finally write

$$F(p) \cdot G(p) = \int_0^\infty \left[\int_0^x e^{-px} f(x-t) g(t) \, dt \right] dx$$

$$= \int_0^\infty e^{-px} \left[\int_0^x f(x-t) g(t) \, dt \right] dx$$

$$= L \left[\int_0^x f(x-t) g(t) \, dt \right].$$

We call the expression $\int_0^x f(x-t) g(t) \, dt$ the *convolution* of f and g. Many texts write

$$f * g(x) = \int_0^x f(x-t) g(t) \, dt. \tag{1}$$

Our calculation shows that

$$L[f * g](p) = F \cdot G = L[f] \cdot L[g].$$

The convolution formula is particularly useful in calculating inverse Laplace transforms.

 EXAMPLE 6.9
Calculate

$$L^{-1} \left[\frac{1}{p^2(p^2+1)} \right].$$

SOLUTION
We write

$$L^{-1} \left[\frac{1}{p^2(p^2+1)} \right] = L^{-1} \left[\frac{1}{p^2} \cdot \frac{1}{p^2+1} \right]$$

$$= \int_0^x (x-t) \cdot \sin t \, dt.$$

Notice that we have recognized that $1/p^2$ is the Laplace transform of x and $1/(p^2+1)$ is the Laplace transform of $\sin x$, and then applied the convolution result.

Now the last integral is easily evaluated (just integrate by parts) and seen to equal

$$x - \sin x.$$

We have thus discovered, rather painlessly, that

$$L^{-1}\left[\frac{1}{p^2(p^2+1)}\right] = x - \sin x.$$

Math Note: All of Fourier and harmonic analysis has versions of the convolution equation that we just described. As an example, suppose that f and g are functions on the interval $[-\pi, \pi]$. Define

$$f * g(x) = \int_{-\pi}^{\pi} f(x - t)g(t)\, dt,$$

where arithmetic is taken to be modulo 2π as usual. Then we can calculate the Fourier coefficients of $f * g$ and it turns out that they are, in a suitable sense, a product of the Fourier coefficients of f and g. We leave the details for you.

An entire area of mathematics is devoted to the study of integral equations of the form

$$f(x) = y(x) + \int_0^x k(x - t)y(t)\, dt. \tag{2}$$

Here f is a given forcing function, and k is a given function known as the *kernel*. Usually k is a mathematical model for the physical process being studied. The objective is to solve for y. As you can see, the integral equation involves a convolution. And, not surprisingly, the Laplace transform comes to our aid in unraveling the equation.

In fact we apply the Laplace transform to both sides of (2). The result is

$$L[f] = L[y] + L[k] \cdot L[y],$$

hence

$$L[y] = \frac{L[f]}{1 + L[k]}.$$

Let us look at an example in which this paradigm occurs.

EXAMPLE 6.10
Use the Laplace transform to solve the integral equation

$$y(x) = x^3 + \int_0^x \sin(x - t)y(t)\, dt.$$

SOLUTION

We apply the Laplace transform to both sides:

$$L[y] = L[x^3] + L[\sin x] \cdot L[y].$$

Solving for $L[y]$, we see that

$$L[y] = \frac{L[x^3]}{1 - L[\sin x]} = \frac{3!/p^4}{1 - 1/(p^2 + 1)}.$$

We may simplify the right-hand side to obtain

$$L[y] = \frac{3!}{p^4} + \frac{3!}{p^6}.$$

Of course it is easy to determine the inverse Laplace transform of the right-hand side. The result is

$$y(x) = x^3 + \frac{x^5}{20}.$$

☞ **You Try It:**   Use the Laplace transform to solve the integral equation

$$y(x) = x^2 + \int_0^x \cos(x - t)y(t)\, dt.$$

We now study an old problem from mechanics that goes back to Niels Henrik Abel (1802–1829). Imagine a wire bent into a smooth curve (Fig. 6.1). The curve terminates at the origin. Imagine a bead sliding from the top of the wire, without friction, down to the origin. The only force acting on the bead is gravity, depending

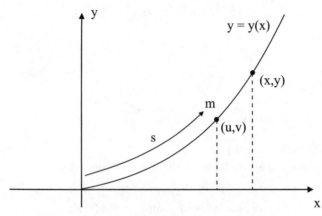

Fig. 6.1.

only on the weight of the bead. Say that the wire is the graph of a function $y = y(x)$. Then the total time for the descent of the bead is some number $T(y)$ that depends on the shape of the wire and on the initial height y. Abel's problem is to run the process in reverse: Suppose that we are given a function T. Then find the shape y of a wire that will result in this time-of-descent function T.

What is interesting about this problem, from the point of view of the present section, is that its mathematical formulation leads to an integral equation of the sort that we have just been discussing. And we will be able to solve it using the Laplace transform.

We begin our analysis with the principle of conservation of energy, namely,

$$\frac{1}{2}m\left(\frac{ds}{dt}\right)^2 = m \cdot g \cdot (y - v).$$

In this equation, m is the mass of the bead, ds/dt is its velocity, and g is the acceleration due to gravity. We use (u, v) as the coordinates of any intermediate point on the curve. The expression on the left-hand side is the standard one from physics for kinetic energy. And the expression on the right is the potential energy.

We may rewrite the last equation as

$$-\frac{ds}{dt} = \sqrt{2g(y - v)}$$

or

$$dt = -\frac{ds}{\sqrt{2g(y - v)}}.$$

Integrating from $v = y$ to $v = 0$ yields

$$T(y) = \int_{v=y}^{v=0} dt = \int_{v=0}^{v=y} \frac{ds}{\sqrt{2g(y - v)}} = \frac{1}{\sqrt{2g}} \int_0^y \frac{s'(v)\, dv}{\sqrt{y - v}}. \tag{3}$$

Now we know from calculus how to calculate the length of a curve:

$$s = s(y) = \int_0^y \sqrt{1 + \left(\frac{dx}{dy}\right)^2}\, dy,$$

hence

$$f(y) = s'(y) = \sqrt{1 + \left(\frac{dx}{dy}\right)^2}. \tag{4}$$

Substituting this last expression into (3), we find that

$$T(y) = \frac{1}{\sqrt{2g}} \int_0^y \frac{f(v)\, dv}{\sqrt{y - v}}. \tag{5}$$

This formula, in principle, allows us to calculate the total descent time $T(y)$ whenever the curve y is given. From the point of view of Abel's problem, the function $T(y)$ is given, and we wish to find y. We think of $f(y)$ as the unknown. The equation (5) is called *Abel's integral equation*.

We note that the integral on the right-hand side of Abel's equation is a convolution (of the functions $y^{-1/2}$ and f). Thus when we apply the Laplace transform to (5) we obtain

$$L[T(y)] = \frac{1}{\sqrt{2g}} L[y^{-1/2}] \cdot L[f(y)].$$

Now we know from Example 6.6 that $L[y^{-1/2}] = \sqrt{\pi/p}$. Hence the last equation may be written as

$$L[f(y)] = \sqrt{2g} \cdot \frac{L[T(y)]}{\sqrt{\pi/p}} = \sqrt{\frac{2g}{\pi}} \cdot p^{1/2} \cdot L[T(y)]. \tag{6}$$

When $T(y)$ is given, then the right-hand side of (6) is completely known, so we can then determine $L[f(y)]$ and hence y (by solving the differential equation (4)).

 EXAMPLE 6.11

Analyze the case of Abel's mechanical problem when $T(y) = T_0$, a constant.

SOLUTION

Our hypothesis means that the time of descent is independent of where on the curve we release the bead. A curve with this property (if in fact one exists) is called a *tautochrone*. In this case the equation (6) becomes

$$L[f(y)] = \sqrt{\frac{2g}{\pi}} p^{1/2} L[T_0] = \sqrt{\frac{2g}{\pi}} p^{1/2} \frac{T_0}{p} = b^{1/2} \cdot \sqrt{\frac{\pi}{p}},$$

where we have used the shorthand $b = 2gT_0^2/\pi^2$. Now $L^{-1}[\sqrt{\pi/p}] = y^{-1/2}$, hence we find that

$$f(y) = \sqrt{\frac{b}{y}}. \tag{7}$$

Now the differential equation (4) tells us that

$$1 + \left(\frac{dx}{dy}\right)^2 = \frac{b}{y},$$

hence

$$x = \int \sqrt{\frac{b-y}{y}}\, dy.$$

Using the change of variable $y = b\sin^2\phi$, we obtain

$$x = 2b \int \cos^2\phi\, d\phi$$

$$= b \int (1 + \cos 2\phi)\, d\phi$$

$$= \frac{b}{2}(2\phi + \sin 2\phi) + C.$$

In conclusion,

$$x = \frac{b}{2}(2\phi + \sin 2\phi) + C \qquad \text{and} \qquad y = \frac{b}{2}(1 - \cos 2\phi). \tag{8}$$

The curve must, by the initial mandate, pass through the origin. Hence $C = 0$. If we put $a = b/2$ and $\theta = 2\phi$, then (8) takes the simpler form

$$x = a(\theta + \sin\theta) \qquad \text{and} \qquad y = a(1 - \cos\theta).$$

These are the parametric equations of a cycloid (Fig. 6.2). A cycloid is a curve generated by a fixed point on the edge of a disc of radius a rolling along the x-axis. See Fig. 6.3. We invite you to work from this synthetic definition to the parametric equations that we just enunciated.

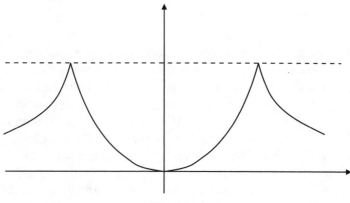

Fig. 6.2.

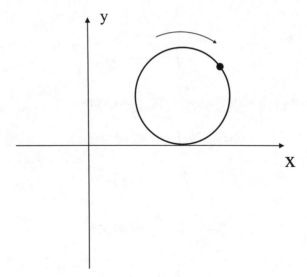

Fig. 6.3.

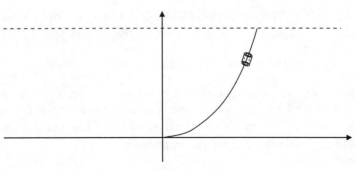

Fig. 6.4.

Math Note: We see that the tautochrone turns out to be a cycloid. This problem and its solution is one of the great triumphs of modern mechanics. An additional very interesting property of this curve is that it is the *brachistochrone*. That means that, given two points *A* and *B* in space, the curve connecting them down which a bead will slide the *fastest* is the cycloid (Fig. 6.4). This last assertion was proved by Isaac Newton, who read the problem as posed by Bernoulli in a periodical. Newton had just come home from a long day at the British Mint (where he served as Director after he gave up his scientific work). He solved the problem in a few hours, and submitted his solution anonymously. But Bernoulli said he knew it was Newton; he "recognized the lion by his claw."

6.5 The Unit Step and Impulse Functions

In this section our goal is to apply the formula

$$L[f * g] = L[f] \cdot L[g]$$

to study the response of an electrical or mechanical system.

Any physical system that responds to a stimulus can be thought of as a device (or black box) that transforms an *input function* (the stimulus) into an *output function* (the response). If we assume that all initial conditions are zero at the moment $t = 0$ when the input f begins to act, then we may hope to solve the resulting differential equation by application of the Laplace transform.

To be more specific, let us consider solutions of the equation

$$y'' + ay' + by = f$$

satisfying the initial conditions $y(0) = 0$ and $y'(0) = 0$. Notice that, since the equation is inhomogeneous, these zero initial conditions cannot force the solution to be identically zero. The input f can be thought of as an impressed external force F or electromotive force E that begins to act at time $t = 0$—just as we discussed when we considered forced vibrations.

When the input function happens to be the unit *step function*

$$u(t) = \begin{cases} 0 & \text{if } t < 0 \\ 1 & \text{if } t \geq 0, \end{cases}$$

then the solution $y(t)$ is denoted by $A(t)$ and is called the *indicial response*. That is to say,

$$A'' + aA' + bA = u. \tag{1}$$

Now, applying the Laplace transform to both sides of (1), and using our standard formulas for the Laplace transforms of derivatives, we find that

$$p^2 L[A] + ap L[A] + b L[A] = L[u] = \frac{1}{p}.$$

So we may solve for $L[A]$ and obtain that

$$L[A] = \frac{1}{p} \cdot \frac{1}{p^2 + ap + b} = \frac{1}{p} \cdot \frac{1}{z(p)}, \tag{2}$$

where $z(p) = p^2 + ap + b$.

Note that we have just been examining the special case of our differential equation with a step function on the right-hand side. Now let us consider the equation in its general form (with an arbitrary external force function):

$$y'' + ay' + by = f.$$

Applying the Laplace transform to both sides (and using our zero initial conditions) gives

$$p^2 L[y] + apL[y] + bL[y] = L[f]$$

or

$$L[y] \cdot z(p) = L[f],$$

so

$$L[y] = \frac{L[f]}{z(p)}. \tag{3}$$

We divide both sides of (3) by p and use (2). The result is

$$\frac{1}{p} \cdot L[y] = \frac{1}{pz(p)} \cdot L[f] = L[A] \cdot L[f].$$

This suggests the use of the convolution theorem:

$$\frac{1}{p} \cdot L[y] = L[A * f].$$

As a result,

$$L[y] = p \cdot L \left[\int_0^t A(t - \tau) f(\tau) \, d\tau \right]$$

$$= L \left[\frac{d}{dt} \int_0^t A(t - \tau) f(\tau) \, d\tau \right].$$

Thus we finally obtain that

$$y(t) = \frac{d}{dt} \int_0^t A(t - \tau) f(\tau) \, d\tau. \tag{4}$$

What we see here is that, once we find the solution A of the differential equation with a step function as an input, then we can obtain the solution for any other input f by convolving A with f and then taking the derivative. With some effort, we can rewrite the equation (4) in an even more appealing way.

In fact we can go ahead and perform the differentiation in (4) to obtain

$$y(t) = \int_0^t A'(t-\tau)f(\tau)\,d\tau + A(0)f(t). \qquad (5)$$

Alternatively, we can use a change of variable to write the convolution as

$$\int_0^t f(t-\sigma)A(\sigma)\,d\sigma.$$

This results in the formula

$$y(t) = \int_0^t f'(t-\sigma)A(\sigma)\,d\sigma + f(0)A(t).$$

Changing variables back again, this gives

$$y(t) = \int_0^t A(t-\tau)f'(\tau)\,d\tau + f(0)A(t).$$

We notice that the initial conditions force $A(0) = 0$ so our other formula (5) becomes

$$y(t) = \int_0^t A'(t-\tau)f(\tau)\,d\tau.$$

Either of these last two displayed formulas is commonly called the *principle of superposition*. They allow us to represent a solution of our differential equation for a general input function in terms of a solution for a step function.

EXAMPLE 6.12
Use the principle of superposition to solve the equation

$$y'' + y' - 6y = 2e^{3t}$$

with initial conditions $y(0) = 0$, $y'(0) = 0$.

SOLUTION
We first observe that

$$z(p) = p^2 + p - 6.$$

Hence

$$L[A] = \frac{1}{p(p^2 + p - 6)}.$$

Now it is a simple matter to apply partial fractions and elementary Laplace transform inversion to obtain

$$A(t) = -\tfrac{1}{6} + \tfrac{1}{15}e^{-3t} + \tfrac{1}{10}e^{2t}.$$

Now $f(t) = 2e^{3t}$, $f'(t) = 6e^{3t}$, and $f(0) = 2$. Thus our first superposition formula gives

$$y(t) = \int_0^t \left[-\tfrac{1}{6} + \tfrac{1}{15}e^{-3(t-\tau)} + \tfrac{1}{10}e^{2(t-\tau)} \right] d\tau$$

$$+ 2\left[-\tfrac{1}{6} + \tfrac{1}{15}e^{-3t} + \tfrac{1}{10}e^{2t} \right]$$

$$= \tfrac{1}{3}e^{3t} + \tfrac{1}{15}e^{-3t} - \tfrac{2}{5}e^{2t}.$$

We invite you to confirm that this is indeed a solution to our initial value problem.

☞ **You Try It:**   Use the principle of superposition to solve the equation

$$y'' + y' + y = 2\cos t$$

with initial conditions $y(0) = 1$, $y'(0) = 0$.

We can use the second principle of superposition, rather than the first, to solve the differential equation. The process is expedited if we first rewrite the equation in terms of an impulse (rather than a step) function.

What is an impulse function? The physicists think of an impulse function as one that takes the value 0 at all points except the origin; at the origin the impulse function takes the value $+\infty$. See Fig. 6.5. In practice, we mathematicians think of an impulse function as a limit of functions

$$\varphi_\epsilon(x) = \begin{cases} 1/\epsilon & \text{if } 0 \le x \le \epsilon \\ 0 & \text{if } x > \epsilon \end{cases}$$

as $\epsilon \to 0^+$ (Fig. 6.6). Observe that, for any $\epsilon > 0$, $\int_0^\infty \varphi_\epsilon(x)\,dx = 1$. It is straightforward to calculate that

$$L[\varphi_\epsilon] = \frac{1 - e^{-p\epsilon}}{p\epsilon}$$

and hence that

$$\lim_{\epsilon \to 0} L[\varphi_\epsilon] \equiv 1.$$

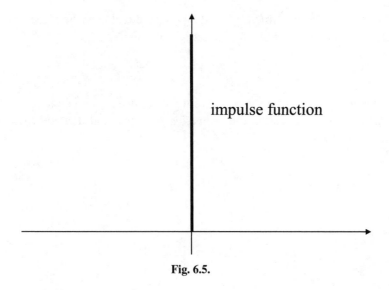

impulse function

Fig. 6.5.

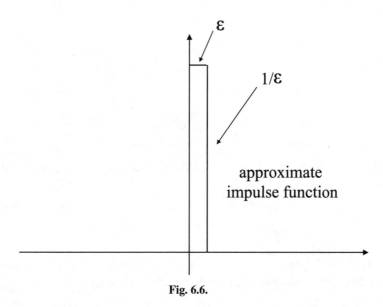

ε

$1/\varepsilon$

approximate
impulse function

Fig. 6.6.

Thus we think of the impulse—intuitively—as an infinitely tall spike at the origin with Laplace transform identically equal to 1. The mathematical justification for the concept of the impulse was outlined in the previous paragraph. A truly rigorous treatment of the impulse requires the theory of distributions (or generalized functions) and we cannot cover it here. It is common to denote the impulse function by $\delta(t)$ (in honor of Paul Dirac (1902–1984), who developed the idea), and to call it the "Dirac function" or "Dirac delta mass." We have that

$$L[\delta] \equiv 1.$$

In the special case that the input function for our differential equation is $f(t) = \delta$, then the solution y is called the *impulsive response* and denoted $h(t)$. In this circumstance we have

$$L[h] = \frac{1}{z(p)},$$

hence

$$h(t) = L^{-1}\left[\frac{1}{z(p)}\right].$$

Now we know that

$$L[A] = \frac{1}{p} \cdot \frac{1}{z(p)} = \frac{L[h]}{p}.$$

As a result,

$$A(t) = \int_0^t h(\tau)\,d\tau.$$

But this last formula shows that $A'(t) = h(t)$, so that our second superposition formula becomes

$$y(t) = \int_0^t h(t - \tau) f(\tau)\,d\tau. \tag{6}$$

In summary, the solution of our differential equation with general input function f is given by the convolution of the impulsive response function with f.

 EXAMPLE 6.13
Solve the differential equation

$$y'' + y' - 6y = 2e^{3t}$$

with initial conditions $y(0) = 0$ and $y'(0) = 0$ using the *second* of our superposition formulas, as rewritten in (6).

SOLUTION

We know that

$$h(t) = L^{-1}\left[\frac{1}{z(p)}\right]$$

$$= L^{-1}\left[\frac{1}{(p+3)(p-2)}\right]$$

$$= L^{-1}\left[\frac{1}{5}\left(\frac{1}{p-2} - \frac{1}{p+3}\right)\right]$$

$$= \frac{1}{5}\left(e^{2t} - e^{-3t}\right).$$

As a result,

$$y(t) = \int_0^t \frac{1}{5}\left[e^{2(t-\tau)} - e^{-3(t-\tau)}\right] 2e^{3t}\, d\tau$$

$$= \frac{1}{3}e^{3t} + \frac{1}{15}e^{-3t} - \frac{2}{5}e^{2t}.$$

Of course this is the same solution that we obtained in the last example, using the other superposition formula.

You Try It: Solve the differential equation

$$y'' + y' + y = 3\sin t$$

with initial conditions $y(0) = 1$ and $y'(0) = 0$ using the *second* of our superposition formulas, as rewritten in (6).

Math Note: To form a more general view of the meaning of convolution, consider a linear physical system in which the effect at the present moment of a small stimulus $g(\tau)\, d\tau$ at any past time τ is proportional to the size of the stimulus. We further assume that the proportionality factor depends only on the elapsed time $t - \tau$, and thus has the form $f(t - \tau)$. The effect at the present time t is therefore

$$f(t - \tau) \cdot g(\tau)\, d\tau.$$

Since the system is linear, the total effect at the present time t due to the stimulus acting throughout the entire past history of the system is obtained by adding these separate effects, and this observation leads to the convolution integral

$$\int_0^t f(t - \tau)g(\tau)\, d\tau.$$

The lower limit is 0 just because we assume that the stimulus started acting at time $t = 0$, i.e., that $g(\tau) = 0$ for all $\tau < 0$. Convolution plays a vital role in the study of wave motion, heat conduction, diffusion, and many other areas of mathematical physics.

Convolutions are important throughout mathematical analysis because they can be used to model translation-invariant processes. You learn more about this idea when you take an advanced course in engineering mathematics, or in Fourier analysis.

Exercises

1. Evaluate the Laplace transform integrals for the third, fourth, and fifth entries in Table 6.1.

2. Without actually integrating, show that
   (a) $L[\sinh ax] = \dfrac{a}{p^2 - a^2}$
   (b) $L[\cosh ax] = \dfrac{p}{p^2 - a^2}$

3. Use the formulas given in the text to find the Laplace transform of each of the following functions:
   (a) 10
   (b) $x^5 + \cos 2x$
   (c) $2e^{3x} - 4 \sin 5x$
   (d) $4 \sin x \cos x + 2e^{-x}$
   (e) $x^6 \sin^2 3x + x^6 \cos^2 3x$

4. Find the Laplace transforms of
   (a) $x^5 e^{-2x}$
   (b) $(1 - x^2)e^{-x}$
   (c) $e^{3x} \cos 2x$

5. Find the inverse Laplace transform of
   (a) $\dfrac{6}{(p + 2)^2 + 9}$
   (b) $\dfrac{12}{(p + 3)^4}$
   (c) $\dfrac{p + 3}{p^2 + 2p + 5}$

6. Solve each of the following differential equations with initial values using the Laplace transform:

   (a) $y' + y = e^{2x}$, $\qquad\qquad$ $y(0) = 0$

   (b) $y'' - 4y' + 4y = 0$, $\qquad$ $y(0) = 0$ and $y'(0) = 3$

   (c) $y'' + 2y' + 2y = 2$, $\qquad$ $y(0) = 0$ and $y'(0) = 1$

7. Solve the initial value problem

$$y' + 4y + 5 \int_0^x y\,dx = e^{-x}, \quad y(0) = 0.$$

8. Calculate each of the following Laplace transforms:

   (a) $L[x^2 \sin ax]$

   (b) $L[xe^x]$

9. Solve each of the following integral equations:

   (a) $y(x) = 1 - \int_0^x (x - t)y(t)\,dt$

   (b) $y(x) = e^x \left[1 + \int_0^x e^{-t}y(t)\,dt \right]$

10. Find the convolution of each of the following pairs of functions:

    (a) $1,\ \sin at$

    (b) $e^{at},\ e^{bt}$ for $a \neq b$

11. Use the method of Laplace transforms to find the general solution of the differential equation (*Hint:* Use the boundary conditions $y(0) = A$ and $y'(0) = B$ to introduce the two undetermined constants that you need):

$$y'' - 5y' + 4y = 0.$$

12. Express this function using one or more step functions, and then calculate the Laplace transform:

$$g(t) = \begin{cases} 0 & \text{if } 0 < t < 3 \\ t - 1 & \text{if } 3 \leq t < \infty. \end{cases}$$

CHAPTER

Numerical Methods

The presentation in this book, or in any standard introductory text on differential equations, can be misleading. A casual reading might lead you to think that "most" differential equations can be solved explicitly, with the solution given by a formula. Such is not the case. Although it can be proved abstractly that almost any ordinary differential equation has a solution—at least locally—it is in general quite difficult to say in any explicit manner what the solution might be. It is sometimes possible to say something qualitative about solutions. And we have also seen that certain important equations that come from physics are fortuitously simple, and can be attacked effectively. But the bottom line is that many of the equations that we *must* solve for engineering or other applications simply do not have closed-form solutions. Just as an instance, the equations that govern the shape of an airplane wing cannot be solved explicitly. Yet we fly every day. How do we come to terms with the intractability of differential equations?

The advent of high-speed digital computers has made it both feasible and, indeed, easy to perform numerical approximation of solutions. The subject of the numerical solution of differential equations is a highly developed one, and is applied daily to problems in engineering, physics, biology, astronomy, and many other parts of science. Solutions may generally be obtained to any desired degree of accuracy, graphs drawn, and almost any necessary analysis performed.

Not surprisingly—and like many of the other fundamental ideas related to calculus—the basic techniques for the numerical solution of differential equations go back to Newton and Euler. This is quite amazing, for these men had no notion of the computing equipment that we have available today. Their insights were quite prescient and powerful.

In the present chapter, we shall only introduce the most basic ideas in the subject of numerical analysis of differential equations. We refer you to [GER], [HIL], [ISK], [STA], and [TOD] for further development of the subject.

7.1 Introductory Remarks

When we create a numerical or discrete model for a differential equation, we make several decisive replacements or substitutions. First, the derivatives in the equation are replaced by *differences* (as in replacing the derivative by a difference quotient). Second, the continuous variable x is replaced by a discrete variable. Third, the real number line is replaced by a discrete set of values. Any type of approximation argument involves some sort of loss of information; that is to say, there will always be an *error term*. It is also the case that these numerical approximation techniques can give rise to instability phenomena and other unpredictable behavior.

The practical significance of these remarks is that numerical methods should never be used in isolation. Whenever possible, the user should also employ qualitative techniques. Endeavor to determine whether the solution is bounded, periodic, or stable. What are its asymptotics at infinity? How do the different solutions interact with each other? In this way you are not using the computing machine blindly, but are instead using the machine to aid and augment your understanding.

The *spirit* of the numerical method is this. Consider the simple differential equation

$$y' = y, \quad y(0) = 1.$$

The initial condition tells us that the point $(0, 1)$ lies on the graph of the solution y. The equation itself tells us that, at that point, the slope of the solution is

$$y' = y = 1.$$

Thus the graph will proceed to the right with slope 1. Let us assume that we shall do our numerical calculation with mesh 0.1. So we proceed to the right to the point $(0.1, 1.1)$. This is the second point on our "approximate solution graph."

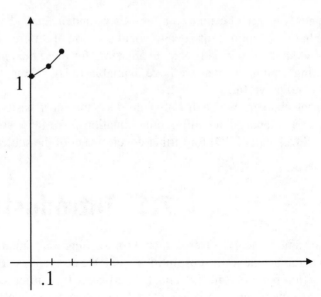

Fig. 7.1.

Now we return to the differential equation to obtain the slope of the solution at this new point. It is

$$y' = y = 1.1.$$

Thus, when we proceed to sketch our approximate solution graph to the right of (0.1, 1.1), we draw a line segment of slope 1.1 to the point (0.2, 1.21), and so forth. See Fig. 7.1.

Of course this is a very simple-minded example, and it is easy to imagine that the approximate solution is diverging rather drastically and unpredictably with each iteration of the method. In subsequent sections we shall learn techniques of Euler (which formalize the method just described) and Runge–Kutta (which give much better, and more reliable, results).

7.2 The Method of Euler

Consider an initial value problem of the form

$$y' = f(x, y), \quad y(x_0) = y_0.$$

We may integrate from x_0 to $x_1 = x_0 + h$ to obtain

$$y(x_1) - y(x_0) = \int_{x_0}^{x_1} f(x, y)\, dx$$

or

$$y(x_1) = y(x_0) + \int_{x_0}^{x_1} f(x, y)\, dx.$$

Since the unknown function y occurs in the integrand on the right, we cannot proceed unless we have some method of approximating the integral.

The Euler method is obtained from the most simple technique for approximating the integral. Namely, we assume that the integrand does not vary much on the interval $[x_0, x_1]$, and therefore that a rather small error will result if we replace $f(x, y)$ by its value at the left endpoint. To wit, we put in place a partition $a = x_0 < x_1 < x_2 < \cdots < x_k = b$ of the interval $[a, b]$ under study. We set $y_0 = y(x_0)$. Now we take

$$y(x_1) = y(x_0) + \int_{x_0}^{x_1} f(x, y)\, dx$$

$$\approx y(x_0) + \int_{x_0}^{x_1} f(x_0, y_0)\, dx$$

$$= y(x_0) + h \cdot f(x_0, y_0).$$

Based on this calculation, we simply *define*

$$y_1 = y_0 + h \cdot f(x_0, y_0).$$

Continuing in this fashion, we set $x_k = x_{k-1} + h$ and define

$$y_{k+1} = y_k + h \cdot f(x_k, y_k).$$

Then the points $(x_0, y_0), (x_1, y_1), \ldots, (x_k, y_k), \ldots$ are the points of our "approximate solution" to the differential equation. Figure 7.2 illustrates the exact solution, the approximate solution, and how they might deviate.

It is sometimes convenient to measure the *total relative error* $\overline{E}_n$ at the nth step; this quantity is defined to be

$$\overline{E}_n = \frac{|y(x_n) - y_n|}{|y(x_n)|}.$$

We usually express this quantity as a percentage, and we obtain thereby a comfortable way of measuring how well the numerical technique under consideration is performing.

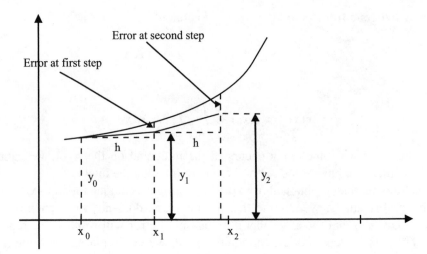

Error at second step

Error at first step

Fig. 7.2.

 EXAMPLE 7.1
Apply the Euler technique to the ordinary differential equation

$$y' = x + y, \quad y(0) = 1 \tag{1}$$

using increments of size $h = 0.2$ and $h = 0.1$.

SOLUTION
We exhibit the calculations in Table 7.1. In the first line of this table, the initial
condition $y(0) = 1$ determines the slope $y' = x + y = 1.00$. Since $h = 0.2$ and
$y_1 = y_0 + h \cdot f(x_0, y_0)$, the next value is given by $1.00 + 0.2 \cdot (1.00) = 1.20$. This
process is iterated in the succeeding lines. We shall retain five decimal places
in this and succeeding tables.

Table 7.1 Tabulated values for exact and numerical solutions to
equation (1) with $h = 0.2$

x_n	y_n	**Exact**	$\overline{E}_n$ (%)
0.0	1.00000	1.00000	0.0
0.2	1.20000	1.24281	3.4
0.4	1.48000	1.58365	6.5
0.6	1.85600	2.04424	9.2
0.8	2.34720	2.65108	11.5
1.0	2.97664	3.43656	13.4

Table 7.2   Tabulated values for exact and numerical solutions to equation (1) with $h = 0.1$

x_n	y_n	Exact	$\overline{E}_n$ (%)
0.0	1.00000	1.00000	0.0
0.1	1.10000	1.11034	0.9
0.2	1.28200	1.24281	1.8
0.3	1.36200	1.39972	2.7
0.4	1.52820	1.58365	3.5
0.5	1.72102	1.79744	4.3
0.6	1.94312	2.04424	4.9
0.7	2.19743	2.32751	5.6
0.8	2.48718	2.65108	6.2
0.9	2.81590	3.01921	6.7
1.0	3.18748	3.43656	7.2

For comparison purposes, we also record in Table 7.2 the tabulated values for $h = 0.1$.

You Try It: Apply the Euler technique to the ordinary differential equation

$$y' = 3x - y, \quad y(0) = 2$$

using increments of size $h = 0.1$.

The displayed data make clear that reducing the step size will increase accuracy. But the tradeoff is that significantly more computation is required. In the next section we shall discuss errors, and in particular at what point there is no advantage to reducing the step size.

Math Note: In calculus class you learned to approximate the value of a function f at a point $x + h$ by $f(x) + h \cdot f'(x)$. For example, try approximating $\sqrt{4.1}$ by letting $f(x) = \sqrt{x}$, $x = 4$, and $h = 0.1$. How can you determine in advance the size of the error in such a calculation?

7.3 The Error Term

The notion of error is central to any numerical technique. Numerical methods only give *approximate answers*. In order for the approximate answer to be useful, we must know how close to the true answer our approximate answer is. Since the whole reason why we went after an approximate answer in the first place was that we had no method for finding the exact answer, this whole discussion raises tricky questions. How do we get our hands on the error, and how do we estimate it? Any time decimal approximations are used, there is a rounding-off procedure involved. *Round-off error* is another critical phenomenon that we must examine.

EXAMPLE 7.2

Examine the differential equation

$$y' = x + y, \quad y(0) = 1 \tag{1}$$

from the numerical point of view, and consider what happens if the step size h is made too small.

SOLUTION

Suppose that we are working with a computer having ordinary precision—which is eight decimal places. This means that all numerical answers are rounded to eight places.

Let $h = 10^{-10}$, a very small step size indeed (but one that could be required for work in microtechnology). Let $f(x, y) = x + y$. Applying the Euler method and computing the first step, we find that the computer yields

$$y_1 = y_0 + h \cdot f(x_0, y_0) = 1 + 10^{-10} = 1.$$

The last equality may seem rather odd—in fact it appears to be false. But this is how the computer will reason: it rounds to eight decimal places! The same phenomenon will occur with the calculation of y_2. In this situation, we see therefore that the Euler method will produce a constant solution—namely, $y \equiv 1$.

The last example is to be taken quite seriously. It describes what would actually happen if you had a canned piece of software to implement Euler's method, and you actually used it on a computer running in the most standard and familiar computing environment. If you are not aware of the dangers of round-off error, and why such errors occur, then you will be a very confused scientist indeed. One way to address the problem is with *double precision*, which gives 16-place decimal accuracy. Another way is to use a symbol manipulation program like `Mathematica` or `Maple` (in which one can preset any number of decimal places of accuracy).

In the present book, we cannot go very deeply into the subject of round-off error. What is most feasible for us is to acknowledge that round-off error must be dealt with in advance, and we shall assume that we have set up our problem so that round-off error is negligible. We shall instead concentrate our discussions on *discretization error*, which is a problem less contingent on artifacts of the computing environment and more central to the theory.

Math Note: How do we, in practice, check to see whether h is too small, and thus causing round-off error? One commonly used technique is to redo the calculation in double precision (on a computer using one of the standard software packages, this would mean 16-place decimal accuracy instead of the usual 8-place accuracy). If the answer seems to change substantially, then some round-off error is probably present in the regular precision (8-place accuracy) calculation.

The local discretization error at the nth step is defined to be $\epsilon_n = y(x_n) - y_n$. Here $y(x_n)$ is the exact value at x_n of the solution of the differential equation, and y_n is the Euler approximation. In fact we may use Taylor's formula to obtain a useful estimate on this error term. To wit, we may write

$$y(x_0 + h) = y_0 + h \cdot y'(x_0) + \frac{h^2}{2} \cdot y''(\xi),$$

for some value of ξ between x_0 and x. But we know, from the differential equation, that

$$y'(x_0) = f(x_0, y_0).$$

Thus

$$y(x_0 + h) = y_0 + h \cdot f(x_0, y_0) + \frac{h^2}{2} \cdot y''(\xi),$$

so that

$$y(x_1) = y(x_0 + h) = y_0 + h \cdot f(x_0, y_0) + \frac{h^2}{2} \cdot y''(\xi) = y_1 + \frac{h^2}{2} \cdot y''(\xi).$$

We may conclude that

$$\epsilon_1 = \frac{h^2}{2} \cdot y''(\xi).$$

Usually on the interval $[x_0, x_n]$ we may see on *a priori* grounds that y'' is bounded by some constant M. Thus our error estimate takes the form

$$|\epsilon_1| \leq \frac{Mh^2}{2}.$$

More generally, the same calculation shows that

$$|\epsilon_j| \leq \frac{Mh^2}{2}.$$

Such an estimate shows us directly, for instance, that if we decrease the step size from h to $h/2$, then the accuracy is increased by a factor of 4.

Unfortunately, in practice things are not as simple as the last paragraph might suggest. For an error is made at *each step* of the Euler method—or of any numerical method—so we must consider the *total discretization error*. This is just the aggregate of all the errors that occur at all steps of the approximation process.

To get a rough estimate of this quantity, we notice that our Euler scheme iterates in n steps, from x_0 to x_n, in increments of length h. So $h = [x_n - x_0]/n$ or $n = [x_n - x_0]/h$. If we assume that the errors accumulate without any cancellation, then the aggregate error is bounded by

$$|E_n| \leq n \cdot \frac{Mh^2}{2} = (x_n - x_0) \cdot \frac{Mh}{2} \equiv C \cdot h.$$

Here $C = (x_n - x_0) \cdot M$, and $(x_n - x_0)$ is of course the length of the interval under study. Thus, for this problem, C is a universal constant. We see that, for Euler's method, the total discretization error is bounded by a constant times the step size.

 EXAMPLE 7.3

Estimate the discretization error, for a step size of 0.2 and for a step size of 0.1, for the differential equation with initial data given by

$$y' = x + y, \quad y(0) = 1. \tag{2}$$

SOLUTION

In order to get the maximum information about the error, we are going to proceed in a somewhat artificial fashion. Namely, we will use the fact that *we can solve the initial value problem explicitly*: the solution is given by $y = 2e^x - x - 1$. Thus $y'' = 2e^x$. Thus, on the interval $[0, 1]$,

$$|y''| \leq 2e^1 = 2e.$$

Hence

$$|\epsilon_j| \leq \frac{Mh^2}{2} \leq \frac{2eh^2}{2} = eh^2$$

for each j. The total discretization error is then bounded (since we calculate this error by summing about $1/h$ terms) by

$$|E_n| \leq eh. \tag{3}$$

Referring to Table 7.1 in Section 7.2 for incrementing by $h = 0.2$, we see that the total discretization error at $x = 1$ is *actually equal to* 0.46 (rounded to two decimal places). [We calculate this error from the table by subtracting y_n from the exact solution.] The error bound given by (3) is $e \cdot (0.2) \approx 0.54$. Of course the actual error is less than this somewhat crude bound. With $h = 0.1$, the actual error from Table 7.2 is 0.25 while the error bound is $e \cdot (0.1) \approx 0.27$.

You Try It: Estimate the discretization error, for a step size of 0.1, for the
differential equation with initial data given by

$$y' = 3x - y, \quad y(1) = 2.$$

Math Note: In practice, we shall not be able to solve the differential equation
being studied. That is, after all, why we are using numerical techniques and a
computer. So how do we, in practice, determine when h is small enough to achieve
the accuracy we desire? A rough-and-ready method, which is used commonly in
the field, is this: Do the calculation for a given h, then for $h/2$, then for $h/4$, and so
forth. When the distance between two successive calculations is within the desired
tolerance for the problem, then it is quite likely that they both are also within the
desired tolerance of the exact solution.

7.4 An Improved Euler Method

We improve the Euler method by following the logical scheme that we employed
when learning numerical methods of integration in calculus class. Namely, our first
method of numerical integration was to approximate a desired integral by a sum of
areas of rectangles. [This is analogous to the Euler method, where we approximate
the integrand by the constant value at its left endpoint.] Next, in integration theory,
we improved our calculations by approximating by a sum of areas of trapezoids.
That amounts to averaging the values at the two endpoints. This is the philosophy
that we now employ.

Recall that our old equation is

$$y_1 = y_0 + \int_{x_0}^{x_1} f(x, y) \, dx.$$

Our idea for Euler's method was to replace the integrand by $f(x_0, y_0)$. This
generated the iterative scheme of the last section. Now we propose to instead
replace the integrand with $[f(x_0, y_0) + f(x_1, y(x_1))]/2$. Thus we find that

$$y_1 = y_0 + \frac{h}{2}[f(x_0, y_0) + f(x_1, y(x_1))]. \tag{1}$$

The trouble with this proposed equation is that $y(x_1)$ is unknown—just because
we do not know the exact solution y. What we can do instead is to replace $y(x_1)$
by its approximate value as found by the Euler method. Denote this new value by
$z_1 = y_0 + h \cdot f(x_0, y_0)$. Then (1) becomes

$$y_1 = y_0 + \frac{h}{2} \cdot [f(x_0, y_0) + f(x_1, z_1)].$$

You should pause to verify that each quantity on the right-hand side can be calculated from information that we have—*without* knowledge of the exact solution of the differential equation. More generally, our iterative scheme is

$$y_{j+1} = y_j + \frac{h}{2} \cdot [f(x_j, y_j) + f(x_{j+1}, z_{j+1})],$$

where

$$z_{j+1} = y_j + h \cdot f(x_j, y_j)$$

and $j = 0, 1, 2, \ldots$.

This new method, usually called the *improved Euler method* or *Heun's method*, first *predicts* and then *corrects* an estimate for y_j. It is an example of a class of numerical techniques called *predictor–corrector methods*. It is possible, using subtle Taylor series arguments, to show that the local discretization error is

$$\epsilon_j = -y'''(\xi) \cdot \frac{h^3}{12},$$

for some value of ξ between x_0 and x_n. Thus, in particular, the total discretization error is proportional to h^2 (instead of h, as before), so we expect more accuracy for the same step size. Figure 7.3 gives a way to visualize the improved Euler method. First, the point at (x_1, z_1) is predicted using the original Euler method, then this point is used to estimate the slope of the solution curve at x_1. This result is then averaged with the original slope estimate at (x_0, y_0) to make a better prediction of the solution—namely, (x_1, y_1).

 EXAMPLE 7.4

Apply the improved Euler method to the differential equation

$$y' = x + y, \quad y(0) = 1 \tag{2}$$

with step size 0.2 and gauge the improvement in accuracy over the ordinary Euler method used in Examples 7.1 and 7.3.

SOLUTION
We see that

$$z_{k+1} = y_k + 0.2 \cdot f(x_k, y_k) = y_k + 0.2(x_k + y_k)$$

and

$$y_{k+1} = y_k + 0.1[(x_k + y_k) + (x_{k+1} + z_{k+1})].$$

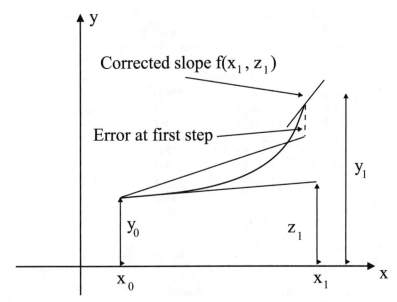

Fig. 7.3.

We begin the calculation by setting $k = 0$ and using the initial values $x_0 = 0.0000$, $y_0 = 1.0000$. Thus

$$z_1 = 1.0000 + 0.2(0.0000 + 1.0000) = 1.2000$$

and

$$y_1 = 1.0000 + 0.1[0.0000 + 1.0000) + (0.2 + 1.2000)] = 1.2400.$$

We continue this process and obtain the values shown in Table 7.3.

Table 7.3 Tabulated values for exact and numerical solutions to (2) with $h = 0.2$ using the improved Euler method

x_n	y_n	Exact	$\overline{E}_n$ (%)
0.0	1.00000	1.00000	0.00
0.2	1.24000	1.24281	0.23
0.4	1.57680	1.58365	0.43
0.6	2.03170	2.04424	0.61
0.8	2.63067	2.65108	0.77
1.0	3.40542	3.43656	0.91

Table 7.4 Tabulated values for exact and numerical solutions to (2) with $h = 0.1$ using the improved Euler method

x_n	y_n	Exact	$\overline{E}_n$ (%)
0.0	1.00000	1.00000	0.0
0.1	1.11000	1.11034	0.0
0.2	1.24205	1.24281	0.1
0.3	1.39847	1.39972	0.1
0.4	1.58180	1.58365	0.1
0.5	1.79489	1.79744	0.1
0.6	2.04086	2.04424	0.2
0.7	2.32315	2.32751	0.2
0.8	2.64558	2.65108	0.2
0.9	3.01236	3.01921	0.2
1.0	3.42816	3.43656	0.2

We see that the resulting approximate value for $y(1)$ is 3.40542. The aggregate error is about 1 percent, whereas with the former Euler method it was more than 13 percent. This is a substantial improvement.

Of course a smaller step size results in even more dramatic improvement in accuracy. Table 7.4 displays the results of applying the improved Euler method to our differential equation using a step size of $h = 0.1$. The relative error at $x = 1.00000$ is now about 0.2 percent, which is another order of magnitude of improvement in accuracy. We have predicted that halving the step size will decrease the aggregate error by a factor of 4. These results bear out that prediction.

 You Try It: Apply the improved Euler method to the differential equation

$$y' = 3x - y, \quad y(1) = 2$$

with step size 0.1 and gauge the improvement in accuracy over the ordinary Euler method.

In the next section we shall use a method of subdividing the intervals of our step sequence to obtain greater accuracy. This results in the Runge–Kutta method.

7.5 The Runge–Kutta Method

Just as the trapezoid rule provides an improvement over the rectangular method for approximating integrals, so Simpson's rule gives an even better means for approximating integrals. With Simpson's rule we approximate not by rectangles or trapezoids but by parabolas.

Check your calculus book (for instance [STE, p. 421]) to review how Simpson's rule works. When we apply it to the integral of f, we find that

$$\int_{x_1}^{x_2} f(x, y)\, dx = \tfrac{1}{6}[f(x_0, y_0) + 4f(x_{1/2}, y(x_{1/2})) + f(x_1, y(x_1))]. \quad (1)$$

Here $x_{1/2} \equiv x_0 + h/2$, the midpoint of x_0 and x_1. We cannot provide all the rigorous details of the derivation of the fourth-order Runge–Kutta method. We instead provide an intuitive development.

Just as we did in obtaining our earlier numerical algorithms, we must now estimate both $y_{1/2}$ and y_1. The first estimate of $y_{1/2}$ comes from Euler's method. Thus

$$y_{1/2} = y_0 + \frac{m_1}{2}.$$

Here

$$m_1 = h \cdot f(x_0, y_0).$$

[The factor of 1/2 here comes from the step size from x_0 to $x_{1/2}$.] To correct the estimate of $y_{1/2}$, we calculate it again in this manner:

$$y_{1/2} = y_0 + \frac{m_2}{2},$$

where

$$m_2 = h \cdot f(x_0 + h/2, y_0 + m_1/2).$$

Now, to predict y_1, we use the expression for $y_{1/2}$ and the Euler method:

$$y_1 = y_{1/2} + \frac{m_3}{2},$$

where $m_3 = h \cdot f(x_0 + h/2, y_0 + m_2/2)$.

Finally, let $m_4 = h \cdot f(x_0 + h, y_0 + m_3)$. The Runge–Kutta scheme is then obtained by substituting each of these estimates into (1) to yield

$$y_1 = y_0 + \tfrac{1}{6}(m_1 + 2m_2 + 2m_3 + m_4).$$

Just as in our earlier work, this algorithm can be applied to any number of mesh points in a natural way. At each step of the iteration, we first compute the four

numbers m_1, m_2, m_3, m_4 given by

$$m_1 = h \cdot f(x_k, y_k),$$

$$m_2 = h \cdot f\left(x_k + \frac{h}{2}, y_k + \frac{m_1}{2}\right),$$

$$m_3 = h \cdot f\left(x_k + \frac{h}{2}, y_k + \frac{m_2}{2}\right),$$

$$m_4 = h \cdot f(x_k + h, y_k + m_3).$$

Then y_{k+1} is given by

$$y_{k+1} = y_k + \tfrac{1}{6}(m_1 + 2m_2 + 2m_3 + m_4).$$

This new analytic paradigm, the Runge–Kutta technique, is capable of giving extremely accurate results without the need for taking very small values of h (thus making the work computationally expensive). The local truncation error is

$$\epsilon_k = -\frac{y^{(5)}(\xi) \cdot h^5}{180},$$

where ξ is a point between x_0 and x_n. The total truncation error is thus of the order of magnitude of h^4.

 EXAMPLE 7.5
Apply the Runge–Kutta method to the differential equation

$$y' = x + y, \quad y(0) = 1. \tag{2}$$

Take $h = 1$, so that the process has only a single step.

SOLUTION
We determine that

$$m_1 = 1 \cdot (0 + 1) = 1,$$

$$m_2 = 1 \cdot (0 + 0.5 + 1 + 0.5) = 2,$$

$$m_3 = 1 \cdot (0 + 0.5 + 1 + 1) = 2.5,$$

$$m_4 = 1 \cdot (0 + 1 + 1 + 2.5) = 4.5.$$

Thus

$$y_1 = 1 + \tfrac{1}{6}(1 + 4 + 5 + 4.5) = 3.417.$$

Table 7.5 Tabulated values for exact and numerical solutions to (2) with $h = 0.2$ using the Runge–Kutta method

x_n	y_n	Exact	$\overline{E}_n$ (%)
0.0	1.00000	1.00000	0.00000
0.2	1.24280	1.24281	0.00044
0.4	1.58364	1.58365	0.00085
0.6	2.04421	2.04424	0.00125
0.8	2.65104	2.65108	0.00152
1.0	3.43650	3.43656	0.00179

Table 7.6 Tabulated values for exact and numerical solutions to (2) with $h = 0.1$ using the Runge–Kutta method

x_n	y_n	Exact	$\overline{E}_n$ (%)
0.0	1.00000	1.00000	0.0
0.1	1.11034	1.11034	0.00002
0.2	1.24281	1.24281	0.00003
0.3	1.39972	1.39972	0.00004
0.4	1.58365	1.58365	0.00006
0.5	1.79744	1.79744	0.00007
0.6	2.04424	2.04424	0.00008
0.7	2.32750	2.32751	0.00009
0.8	2.65108	2.65108	0.00010
0.9	3.01920	3.01921	0.00011
1.0	3.43656	3.43656	0.00012

Observe that this approximate solution is even better than that obtained with the improved Euler method for $h = 0.2$. And the amount of computation involved was absolutely minimal.

Table 7.5 shows the result of applying Runge–Kutta to our differential equation with $h = 0.2$. Notice that our approximate value for $y(1)$ is 3.43650, which agrees with the exact value to four decimal places. The relative error is less than 0.002 percent.

If we cut the step size in half, to 0.1, then the accuracy is increased dramatically—see Table 7.6. Now the relative error is less than 0.0002 percent.

You Try It: Apply the Runge–Kutta method to the differential equation

$$y' = 2x - y, \quad y(1) = 2.$$

Take $h = 0.2$.

Math Note: We refine our methods of estimating integrals by passing from approximation by rectangles to approximation by trapezoids and then to approximation by parabolas. We follow a similar scheme in refining our numerical methods for differential equations. There is nothing to prevent us from continuing these refinements—to approximation by cubics, and then by quartics, and so forth. But there is a tradeoff in that the calculations become very complicated rather quickly, and thus computationally expensive. Most of the modern techniques are refinements of the method of approximation by parabolas.

Exercises

1. In each problem, use the Euler method with $h = 0.1$ to estimate the solution at $x = 1$. In each case, compare your results to the *exact* solution and discuss how well (or poorly) the Euler method has worked.
   (a) $y' = 2x + 2y$, $y(0) = 1$
   (b) $y' = 1/y$, $y(0) = 1$

2. In each problem, use the exact solution, together with step sizes $h = 0.2$, to estimate the total discretization error that occurs with the Euler method at $x = 1$.
   (a) $y' = 2x + 2y$, $y(0) = 1$
   (b) $y' = 1/y$, $y(0) = 1$

3. In each problem, use the improved Euler method with $h = 0.1$ to estimate the solution at $x = 1$. Compare your results to the *exact* solution.
   (a) $y' = 2x + 2y$, $y(0) = 1$
   (b) $y' = 1/y$, $y(0) = 1$

4. In each problem, use the Runge–Kutta method with $h = 0.1$ to estimate the solution at $x = 1$. Compare your results to the *exact* solution.
   (a) $y' = 2x + 2y$, $y(0) = 1$
   (b) $y' = 1/y$, $y(0) = 1$

5. Use the Euler method with $h = 0.01$ to estimate the solution at $x = 0.02$. Compare your result to the *exact* solution and discuss how well (or poorly) the Euler method has worked.

$$y' = x - 2y, \quad y(0) = 2.$$

6. Use the improved Euler method with $h = 0.01$ to estimate the solution at $x = 0.02$. Compare your result to the *exact* solution.

$$y' = x - 2y, \quad y(0) = 2.$$

7. Use the Runge–Kutta method with $h = 0.01$ to estimate the solution at $x = 0.02$. Compare your result to the *exact* solution.

$$y' = x - 2y, \quad y(0) = 2.$$

CHAPTER 8

Systems of First-Order Equations

8.1 Introductory Remarks

Systems of differential equations arise very naturally in many physical contexts. If $y_1, y_2, \ldots, y_n$ are functions of the variable x, then a system, for us, will have the form

$$
\begin{aligned}
y_1' &= f_1(y_1, \ldots, y_n) \\
y_2' &= f_2(y_1, \ldots, y_n) \\
&\cdots \\
y_n' &= f_n(y_1, \ldots, y_n).
\end{aligned}
\tag{1}
$$

In Section 2.7 we used a system of two second-order equations to describe the motion of coupled harmonic oscillators. In an example below we shall see how a system occurs in the context of dynamical systems having several degrees of freedom. In another context, we shall see a system of differential equations used to model a predator–prey problem in the study of population ecology.

From the mathematical point of view, systems of equations are useful in part because an nth-order equation

$$y^{(n)} = f(x, y, y', \ldots, y^{(n-1)}) \tag{2}$$

can be regarded (after a suitable change of notation) as a system. To see this, we let

$$y_0 = y, \quad y_1 = y', \quad \ldots, \quad y_{n-1} = y^{(n-1)}.$$

Then we have

$$y_1' = y_2$$

$$y_2' = y_3$$

$$\ldots$$

$$y_n' = f(x, y_1, y_2, \ldots, y_n),$$

and this system is equivalent to our original equation (2). In practice, it is sometimes possible to treat a system like this as a vector-valued, first-order differential equation, and to use techniques that we have studied in this book to learn about the (vector) solution.

For cultural reasons, and for general interest, we shall next turn to the n-body problem of classical mechanics. It, too, can be modeled by a system of ordinary differential equations. Imagine n particles with masses m_j, $j = 1, \ldots, n$, and located at points (x_j, y_j, z_j) in three-dimensional space. Assume that these points exert a force on each other according to Newton's Law of Universal Gravitation (which we shall formulate in a moment). If r_{ij} is the distance between m_i and m_j and if θ is the angle from the positive x-axis to the segment joining them (Fig. 8.1), then the component of the force exerted on m_i by m_j is

$$\frac{Gm_im_j}{r_{ij}^2} \cos\theta = \frac{Gm_im_j(x_j - x_i)}{r_{ij}^3}.$$

Here G is a constant that depends on the force of gravity. Since the sum of all these components for $i \neq j$ equals $m_i(d^2x_i/dt^2)$ (by Newton's second law), we obtain

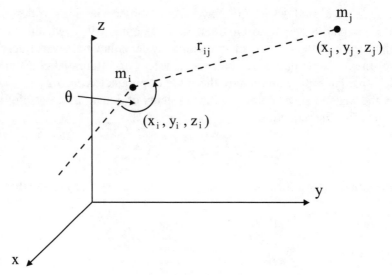

Fig. 8.1.

n second-order differential equations

$$\frac{d^2x_i}{dt^2} = G \cdot \sum_{j \neq i} \frac{m_j(x_j - x_i)}{r_{ij}^3};$$

similarly,

$$\frac{d^2y_i}{dt^2} = G \cdot \sum_{j \neq i} \frac{m_j(y_j - y_i)}{r_{ij}^3}$$

and

$$\frac{d^2z_i}{dt^2} = G \cdot \sum_{j \neq i} \frac{m_j(z_j - z_i)}{r_{ij}^3}.$$

If we make the change of notation

$$v_{x_i} = \frac{dx_i}{dt}, \quad v_{y_i} = \frac{dy_i}{dt}, \quad v_{z_i} = \frac{dz_i}{dt},$$

then we can reduce our system of $3n$ second-order equations to $6n$ first-order equations with unknowns $x_1, v_{x_1}, x_2, v_{x_2}, \ldots, x_n, v_{x_n}, y_1, v_{y_1}, y_2, v_{y_2}, \ldots, y_n, v_{y_n}, z_1, v_{z_1}, z_2, v_{z_2}, \ldots, z_n, v_{z_n}$. We can also make the substitution

$$f_{ij}^3 = \left[(x_i - x_j)^2 + (y_i - y_j)^2 + (z_i - z_j)^2\right]^{3/2}.$$

Then it can be proved that, if initial positions and velocities are specified for each of the n particles and if the particles do not collide (i.e., r_{ij} is never 0), then the subsequent position and velocity of each particle in the system is uniquely determined.

This is the Newtonian model of the universe. It is thoroughly deterministic. If $n = 2$, then the system was completely solved by Newton, giving rise to Kepler's laws (Section 2.6). But for $n \geq 3$ there is a great deal that is not known. Of course this mathematical model can be taken to model the motions of the planets in our solar system. It is not known, for example, whether one of the planets (the Earth, let us say) will one day leave its orbit and go crashing into the sun. Or whether another planet will suddenly leave its orbit and go shooting out to infinity.

8.2 Linear Systems

Our experience in this subject might lead us to believe that systems of *linear* equations will be the most tractable. That is indeed the case; we treat them in this section. By way of introduction, we shall concentrate on systems of two first-order equations in two unknown functions. Thus we have

$$\begin{cases} \dfrac{dx}{dt} = F(t, x, y) \\ \dfrac{dy}{dt} = G(t, x, y). \end{cases}$$

The brace is used here to stress that the equations are linked; the choice of t for the independent variable and of x and y for the dependent variables is traditional and will be borne out in the ensuing discussions.

In fact our system will have an even more special form because of linearity:

$$\begin{cases} \dfrac{dx}{dt} = a_1(t)x + b_1(t)y + f_1(t) \\ \dfrac{dy}{dt} = a_2(t)x + b_2(t)y + f_2(t). \end{cases} \tag{1}$$

It will be convenient, and it is physically natural, for us to assume that the coefficient functions $a_j, b_j, f_j, j = 1, 2$, are continuous on a closed interval $[a, b]$ in the t-axis.

In the special case that $f_1 = f_2 \equiv 0$, then we call the system *homogeneous*. Otherwise it is *nonhomogeneous*. A solution of this system is of course a *pair* of functions $(x(t), y(t))$ that satisfy both differential equations. We shall write

$$\begin{cases} x = x(t) \\ y = y(t). \end{cases}$$

Most of the systems that we shall study in any detail will have constant coefficients.

 EXAMPLE 8.1
Verify that the system

$$\begin{cases} \dfrac{dx}{dt} = 4x - y \\[2mm] \dfrac{dy}{dt} = 2x + y \end{cases}$$

has

$$\begin{cases} x = e^{3t} \\ y = e^{3t} \end{cases}$$

and

$$\begin{cases} x = e^{2t} \\ y = 2e^{2t} \end{cases}$$

as solution sets.

SOLUTION
We shall verify the first solution set, and leave the second for you.

Substituting $x = e^{3t}$, $y = e^{3t}$ into the first equation yields

$$\frac{d}{dt} e^{3t} = 4e^{3t} - e^{3t}$$

or

$$3e^{3t} = 3e^{3t},$$

so that equation checks. For the second equation, we obtain

$$\frac{d}{dt} e^{3t} = 2e^{3t} + e^{3t}$$

or

$$3e^{3t} = 3e^{3t},$$

so the second equation checks.

You Try It: Verify that the system

$$\begin{cases} \dfrac{dx}{dt} = \dfrac{2x+y}{2} \\[2mm] \dfrac{dy}{dt} = 2x + y \end{cases}$$

has

$$\begin{cases} x = e^{2t} \\ y = 2e^{2t} \end{cases}$$

as a solution set.

We now give a sketch of the general theory of linear systems of first-order equations. It is a fact that any second-order linear equation may be reduced to a first-order system. Thus it will not be surprising that the theory we are about to describe is similar to the theory of second-order linear equations.

We begin with a fundamental existence and uniqueness theorem.

THEOREM 8.1

Given

Let $[a, b]$ be an interval and $t_0 \in [a, b]$. Let x_0 and y_0 be arbitrary numbers. Then there is one and only one solution to the system

$$\begin{cases} \dfrac{dx}{dt} = a_1(t)x + b_1(t)y + f_1(t) \\[2mm] \dfrac{dy}{dt} = a_2(t)x + b_2(t)y + f_2(t) \end{cases} \tag{1}$$

satisfying $x(t_0) = x_0$, $y(t_0) = y_0$.

We next discuss the structure of the solution of (1) that is obtained when $f_1(t) = f_2(t) \equiv 0$ (the so-called *homogeneous* situation). Thus we have

$$\begin{cases} \dfrac{dx}{dt} = a_1(t)x + b_1(t)y \\[2mm] \dfrac{dy}{dt} = a_2(t)x + b_2(t)y. \end{cases} \tag{2}$$

Of course the identically zero solution ($x(t) \equiv 0$, $y(t) \equiv 0$) is a solution of this homogeneous system. The next theorem—familiar in form—will be the key to constructing more useful solutions.

Given | **THEOREM 8.2**

If the homogeneous system (2) has two solutions

$$\begin{cases} x = x_1(t) \\ y = y_1(t) \end{cases} \quad and \quad \begin{cases} x = x_2(t) \\ y = y_2(t) \end{cases} \tag{3}$$

on $[a, b]$, *then, for any constants* c_1 *and* c_2,

$$x = c_1 x_1(t) + c_2 x_2(t)$$
$$y = c_1 y_1(t) + c_2 y_2(t) \tag{4}$$

is also a solution on $[a, b]$.

Note, in the last theorem, that a new solution is obtained from the original two by multiplying the first by c_1 and the second by c_2 and then adding. We therefore call the newly created solution a *linear combination* of the given solutions. Thus Theorem 8.2 simply says that a linear combination of two solutions of the homogeneous linear system is also a solution of the system. As an instance, in Example 8.1, any pair of functions of the form

$$\begin{cases} x = c_1 e^{3t} + c_2 e^{2t} \\ y = c_1 e^{3t} + c_2 2 e^{2t} \end{cases} \tag{5}$$

is a solution of the given system.

The next obvious question to settle is whether the collection of all linear combinations of two independent solutions of the homogeneous system is in fact *all* the solutions (i.e., the *general solution*) of the system. By Theorem 8.1, we can generate all possible solutions provided we can arrange to satisfy all possible sets of initial conditions. This will now reduce to a simple and familiar algebra problem.

Demanding that, for some choice of c_1 and c_2, the solution

$$\begin{cases} x = c_1 e^{3t} + c_2 e^{2t} \\ y = c_1 e^{3t} + c_2 e^{2t} \end{cases}$$

satisfy $x(t_0) = x_0$ and $y(t_0) = y_0$ amounts to specifying that

$$x_0 = c_1 x_1(t_0) + c_2 x_2(t_0)$$

and

$$y_0 = c_1 y_1(t_0) + c_2 y_2(t_0).$$

This will be possible, for any choice of x_0 and y_0, provided that the determinant of the coefficients of the linear system not be zero. In other words, we require that

$$W(t) = \det \begin{pmatrix} x_1(t) & x_2(t) \\ y_1(t) & y_2(t) \end{pmatrix} = x_1(t) y_2(t) - y_1(t) x_2(t) \neq 0$$

on the interval $[a, b]$. This determinant is naturally called the *Wronskian* of the two solutions.

Our discussion thus far establishes the following theorem:

THEOREM 8.3

If the two solutions (3) of the homogeneous system (2) have a nonvanishing Wronskian on the interval $[a, b]$, then (4) is the general solution of the system on this interval.

Thus, in particular, (5) is the general solution of the system of differential equations in Example 8.1—for the Wronskian of the two solution sets is

$$W(t) = \det \begin{pmatrix} e^{3t} & e^{2t} \\ e^{3t} & 2e^{2t} \end{pmatrix} = e^{5t},$$

and this function of course never vanishes.

As in our previous applications of the Wronskian (see in particular Section 5.5), it is now still the case that either the Wronskian is identically zero or else it is never vanishing. For the record, we enunciate this property formally.

THEOREM 8.4

If $W(t)$ is the Wronskian of the two solutions of our homogeneous system (2), then either W is identically zero or else it is nowhere vanishing.

Math Note: It is possible to think of a system of differential equations as just a single vector-valued ordinary differential equation $Y'(t) = F(t, Y)$, with $Y = (x, y)$. In many circumstances the solution to such an equation is an exponential, suitably interpreted (just as it is for the scalar-valued differential equations that we studied earlier in the book). We shall not explore this matter here, but some of the references provide details of this approach.

We now develop an alternative approach to the question of whether a given pair of solutions generates the general solution of a system. This new method is often more direct and more convenient.

The two solutions (3) are called *linearly dependent* on the interval $[a, b]$ if one ordered pair (x_1, y_1) is a constant multiple of the other. Thus they are *linearly dependent* if there is a constant k such that

$$\begin{aligned} x_1(t) &= k \cdot x_2(t) \\ y_1(t) &= k \cdot y_2(t) \end{aligned} \qquad \text{or} \qquad \begin{aligned} x_2(t) &= k \cdot x_1(t) \\ y_2(t) &= k \cdot y_1(t) \end{aligned}$$

for some constant k and for all $t \in [a, b]$. The solutions are *linearly independent* if neither is a constant multiple of the other in the sense just indicated. Clearly linear dependence is equivalent to the condition that there exist two constants c_1 and c_2, not both zero, such that

$$c_1 x_1(t) + c_2 x_2(t) = 0$$

$$c_1 y_1(t) + c_2 y_2(t) = 0$$

for all $t \in [a, b]$.

THEOREM 8.5

If the two solutions (2) of the homogeneous system (2) are linearly independent on the interval $[a, b]$, then (4) is the general solution of (2) on this interval.

The interest of this new test is that one can usually determine by inspection whether two solutions are linearly independent.

Now it is time to return to the general case—of nonhomogeneous (or inhomogeneous) systems. We conclude our discussion with this result (and, again, note the analogy with second-order linear equations).

THEOREM 8.6

If the two solutions (3) of the homogeneous system (2) are linearly independent on $[a, b]$ and if

$$\begin{cases} x = x_p(t) \\ y = y_p(t) \end{cases}$$

is any particular solution of the system (1) on this interval, then

$$\begin{cases} x = c_1 x_1(t) + c_2 x_2(t) + x_p(t) \\ y = c_1 y_1(t) + c_2 y_2(t) + y_p(t) \end{cases}$$

is the general solution of (1) on $[a, b]$.

Although we would like to end this section with a dramatic example tying all the ideas together, this is in fact not feasible. In general it is quite difficult to find both a particular solution and the general solution to the associated homogeneous equations for a given system. We shall be able to treat the matter most effectively for systems with constant coefficients. We learn about that situation in the next section.

8.3 Homogeneous Linear Systems with Constant Coefficients

It is now time for us to give a complete and explicit solution of the system

$$\begin{cases} \dfrac{dx}{dt} = a_1 x + b_1 y \\ \dfrac{dy}{dt} = a_2 x + b_2 y. \end{cases} \tag{1}$$

Here a_1, a_2, b_1, b_2 are given constants. Sometimes a system of this type can be solved by differentiating one of the two equations, eliminating one of the dependent variables, and then solving the resulting second-order linear equation. In this section we propose an alternative method that is based on constructing a pair of linearly independent solutions directly from the given system.

Working by analogy with our studies of first-order linear equations, we now posit that our system has a solution of the form

$$\begin{cases} x = A e^{mt} \\ y = B e^{mt}. \end{cases} \tag{2}$$

We substitute (2) into (1) and obtain

$$Am e^{mt} = a_1 A e^{mt} + b_1 B e^{mt}$$

$$Bm e^{mt} = a_2 A e^{mt} + b_2 B e^{mt}.$$

Dividing out the common factor of e^{mt} and rearranging yields the associated linear algebraic system

$$\begin{aligned} (a_1 - m)A + b_1 B &= 0 \\ a_2 A + (b_2 - m)B &= 0 \end{aligned} \tag{3}$$

in the unknowns A and B.

Of course the system (3) has the trivial solution $A = B = 0$. This makes (2) the trivial solution of (1). We are of course seeking nontrivial solutions. The algebraic system (3) will have nontrivial solutions precisely when the determinant of the coefficients vanishes, i.e.,

$$\det \begin{pmatrix} a_1 - m & b_1 \\ a_2 & b_2 - m \end{pmatrix} = 0.$$

Expanding the determinant, we find this quadratic expression for the unknown m:

$$m^2 - (a_1 + b_2)m + (a_1 b_2 - a_2 b_1) = 0. \tag{4}$$

We call this the *associated equation* for the original system (1).

Let m_1, m_2 be the roots of the equation (4). If we replace m by m_1 in (4), then we know that the resulting equations have a nontrivial solution set A_1, B_1 so that

$$\begin{cases} x = A_1 e^{m_1 t} \\ y = B_1 e^{m_1 t} \end{cases} \tag{5}$$

is a nontrivial solution of the original system (1). Proceeding similarly with m_2, we find another nontrivial solution,

$$\begin{cases} x = A_2 e^{m_2 t} \\ y = B_2 e^{m_2 t}. \end{cases} \tag{6}$$

In order to be sure that we obtain two linearly independent solutions, and hence the general solution for (1), we must examine in detail each of the three possibilities for m_1 and m_2.

8.3.1 DISTINCT REAL ROOTS

When m_1 and m_2 are distinct real numbers, then the solutions (5) and (6) are linearly independent. For, in fact, $e^{m_1 t}$ and $e^{m_2 t}$ are linearly independent. Thus

$$\begin{cases} x = c_1 A_1 e^{m_1 t} + c_2 A_2 e^{m_2 t} \\ y = c_1 B_1 e^{m_1 t} + c_2 B_2 e^{m_2 t} \end{cases}$$

is the general solution of (1).

 EXAMPLE 8.2
Find the general solution of the system

$$\begin{cases} \dfrac{dx}{dt} = x + y \\ \dfrac{dy}{dt} = 4x - 2y. \end{cases}$$

SOLUTION
The associated algebraic system is

$$(1 - m)A + B = 0$$
$$4A + (-2 - m)B = 0. \tag{7}$$

The auxiliary equation is then

$$m^2 + m - 6 = 0 \quad \text{or} \quad (m + 3)(m - 2) = 0,$$

so that $m_1 = -3, m_2 = 2$.

With m_1, the set of equations (7) becomes

$$4A + B = 0$$

$$4A + B = 0.$$

Since these equations are identical, it is plain that the determinant of the coefficients is zero and there do exist nontrivial solutions.

A simple nontrivial solution of our system is $A = 1$, $B = -4$. Thus

$$\begin{cases} x = e^{-3t} \\ y = -4e^{-3t} \end{cases}$$

is a nontrivial solution of our original system of differential equations.

With m_2, the set of equations (7) becomes

$$-A + B = 0$$

$$4A - 4B = 0.$$

Plainly these equations are multiples of each other, and there do exist nontrivial solutions.

A simple nontrivial solution of our system is $A = 1$, $B = 1$. Thus

$$\begin{cases} x = e^{2t} \\ y = e^{2t} \end{cases}$$

is a nontrivial solution of our original system of differential equations.

Clearly the two solution sets that we have found are linearly independent. Thus

$$\begin{cases} x = c_1 e^{-3t} + c_2 e^{2t} \\ y = -4c_1 e^{-3t} + c_2 e^{2t} \end{cases}$$

is the general solution of our system.

You Try It:　Find the general solution of the system

$$\begin{cases} \dfrac{dx}{dt} = x - 3y \\ \dfrac{dy}{dt} = x + 2y. \end{cases}$$

8.3.2 DISTINCT COMPLEX ROOTS

In fact the only way that complex roots can occur as roots of a quadratic equation with real coefficients is as distinct conjugate roots $a \pm ib$, where a and b are real numbers and $b \neq 0$. In this case we expect the coefficients A and B to be complex numbers (which, for convenience, we shall call A_j^* and B_j^*), and we obtain the two linearly independent solutions

$$\begin{cases} x = A_1^* e^{(a+ib)t} \\ y = B_1^* e^{(a+ib)t} \end{cases} \quad \text{and} \quad \begin{cases} x = A_2^* e^{(a-ib)t} \\ y = B_2^* e^{(a-ib)t}. \end{cases} \tag{8}$$

However, these are complex-valued solutions. On physical grounds, we often want real-valued solutions; we therefore need a procedure for extracting such solutions.

We write $A_1^* = A_1 + i A_2$ and $B_1^* = B_1 + i B_2$, and we apply Euler's formula to the exponential. Thus the first indicated solution becomes

$$\begin{cases} x = (A_1 + i A_2)e^{at}(\cos bt + i \sin bt) \\ y = (B_1 + i B_2)e^{at}(\cos bt + i \sin bt). \end{cases}$$

We may rewrite this as

$$\begin{cases} x = e^{at}\big[(A_1 \cos bt - A_2 \sin bt) + i(A_1 \sin bt + A_2 \cos bt)\big] \\ y = e^{at}\big[(B_1 \cos bt - B_2 \sin bt) + i(B_1 \sin bt + B_2 \cos bt)\big]. \end{cases}$$

From this information, just as in the case of single differential equations (Section 2.1), we deduce that there are two real-valued solutions to the system:

$$\begin{cases} x = e^{at}(A_1 \cos bt - A_2 \sin bt) \\ y = e^{at}(B_1 \cos bt - B_2 \sin bt) \end{cases} \tag{9}$$

and

$$\begin{cases} x = e^{at}(A_1 \sin bt + A_2 \cos bt) \\ y = e^{at}(B_1 \sin bt + B_2 \cos bt). \end{cases} \tag{10}$$

One can use just algebra to see that these solutions are linearly independent (exercise for you). Thus the general solution to our linear system of ordinary differential equations is

$$\begin{cases} x = e^{at}\big[c_1(A_1 \cos bt - A_2 \sin bt) + c_2(A_1 \sin bt + A_2 \cos bt)\big] \\ y = e^{at}\big[c_1(B_1 \cos bt - B_2 \sin bt) + c_2(B_1 \sin bt + B_2 \cos bt)\big]. \end{cases}$$

Since this already gives us the general solution of our system, there is no need to consider the second of the two solutions given in (8). Just as in the case of a single differential equation of second-order, our analysis of that second solution would give rise to the same general solution.

EXAMPLE 8.3

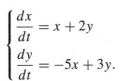

Find the general solution of the system

$$\begin{cases} \dfrac{dx}{dt} = x + 2y \\[2mm] \dfrac{dy}{dt} = -5x + 3y. \end{cases}$$

SOLUTION

The associated algebraic system is

$$(1-m)A + 2B = 0$$
$$-5A + (3-m)B = 0. \tag{7}$$

The auxiliary equation is then

$$m^2 - 4m + 13 = 0.$$

We therefore see that

$$m = \frac{4 \pm \sqrt{4^2 - 4 \cdot 1 \cdot 13}}{2 \cdot 1} = \frac{4 \pm 6i}{2} = 2 \pm 3i.$$

For $m = 2 + 3i$, we solve the system

$$(-1 - 3i)A + 2B = 0$$

$$-5A + (1 - 3i)B = 0$$

and find that $A = 1$, $B = 1/2 + (3/2)i$. Likewise, for $m = 2 - 3i$, we solve the system

$$(-1 + 3i)A + 2B = 0$$

$$-5A + (1 + 3i)B = 0$$

and find that $A = 1$, $B = 1/2 - (3/2)i$. Thus the complex solution sets to our system are

$$x = e^{(2+3i)t}, \quad y = \left(\tfrac{1}{2} + \tfrac{3}{2}i\right) e^{(2+3i)t}$$

and

$$x = e^{(2-3i)t}, \quad y = \left(\tfrac{1}{2} - \tfrac{3}{2}i\right) e^{(2-3i)t}.$$

The real solution sets are then

$$x = e^{2t} \cos 3t, \quad y = \tfrac{1}{2}e^{2t} \cos 3t - \tfrac{3}{2}e^{2t} \sin 3t$$

and

$$x = e^{2t} \sin 3t, \quad y = \tfrac{1}{2}e^{2t} \sin 3t + \tfrac{3}{2}e^{2t} \cos 3t.$$

You Try It: Find the general solution of the system

$$\begin{cases} \dfrac{dx}{dt} = -2x + y \\[2mm] \dfrac{dy}{dt} = -x - 3y. \end{cases}$$

8.3.3 REPEATED REAL ROOTS

When $m_1 = m_2 = m$, then (5) and (6) are not linearly independent; in this case we have just the one solution

$$\begin{cases} x = Ae^{mt} \\ y = Be^{mt}. \end{cases}$$

Our experience with repeated roots of the auxiliary equation in the case of second-order linear equations with constant coefficients might lead us to guess that there is a second solution obtained by introducing into each of x and y a coefficient of t. In fact the present situation calls for something a bit more elaborate. We seek a second solution of the form

$$\begin{cases} x = (A_1 + A_2 t)e^{mt} \\ y = (B_1 + B_2 t)e^{mt}. \end{cases} \tag{11}$$

The general solution is then

$$\begin{cases} x = c_1 Ae^{mt} + c_2(A_1 + A_2 t)e^{mt} \\ y = c_1 Be^{mt} + c_2(B_1 + B_2 t)e^{mt}. \end{cases} \tag{12}$$

The constants A_1, A_2, B_1, B_2 are determined by substituting (11) into the original system of differential equations. Rather than endeavor to carry out this process in complete generality, we now illustrate the idea with a simple example.[1]

---

[1]There is an exception to the general discussion we have just presented that we ought to at least note. Namely, in case the coefficients of the system of ordinary differential equations satisfy $a_1 = b_2 = a$ and $a_2 = b_1 = 0$, then the associated quadratic equation is $m^2 - 2ma + a^2 = (m - a)^2 = 0$. Thus $m = a$ and the constants A and

EXAMPLE 8.4

e.g.

Find the general solution of the system

$$\begin{cases} \dfrac{dx}{dt} = 3x - 4y \\[2mm] \dfrac{dy}{dt} = x - y. \end{cases}$$

SOLUTION

The associated linear algebraic system is

$$(3 - m)A - 4B = 0$$

$$A + (-1 - m)B = 0.$$

The auxiliary quadratic equation is then

$$m^2 - 2m + 1 = 0 \qquad \text{or} \qquad (m - 1)^2 = 0.$$

Thus $m_1 = m_2 = m = 1$.

With $m = 1$, the linear system becomes

$$2A - 4B = 0$$

$$A - 2B = 0.$$

Of course $A = 2$, $B = 1$ is a solution, so we have

$$\begin{cases} x = 2e^t \\ y = e^t \end{cases}$$

as a nontrivial solution of the given system.

We now seek a second linearly independent solution of the form

$$\begin{cases} x = (A_1 + A_2 t)e^t \\ y = (B_1 + B_2 t)e^t. \end{cases} \tag{13}$$

B are completely unrestricted (i.e., the putative equations that we usually solve for A and B reduce to a trivial tautology). In this case the general solution of our system of differential equations is just

$$\begin{cases} x = c_1 e^{mt} \\ y = c_2 e^{mt}. \end{cases}$$

What is going on here is that each differential equation can be solved independently; there is no interdependence. We call such a system *uncoupled*.

When these expressions are substituted into our system of differential equations, we find that

$$(A_1 + A_2 t + A_2)e^t = 3(A_1 + A_2 t)e^t - 4(B_1 + B_2 t)e^t$$

$$(B_1 + B_2 t + B_2)e^t = (A_1 + A_2 t)e^t - (B_1 + B_2 t)e^t.$$

Using a little algebra, these can be reduced to

$$(2A_2 - 4B_2)t + (2A_1 - A_2 - 4B_1) = 0$$

$$(A_2 - 2B_2)t + (A_1 - 2B_1 - B_2) = 0.$$

Since these last are to be identities in the variable t, we can only conclude that

$$2A_2 - 4B_2 = 0 \qquad 2A_1 - A_2 - 4B_1 = 0$$

$$A_2 - 2B_2 = 0 \qquad A_1 - 2B_1 - B_2 = 0.$$

The two equations on the left have $A_2 = 2$, $B_2 = 1$ as a solution. With these values, the two equations on the right become

$$2A_1 - 4B_1 = 2$$

$$A_1 - 2B_1 = 1.$$

Of course their solution is $A_1 = 1$, $B_1 = 0$. We now insert these numbers into (13) to obtain

$$\begin{cases} x = (1 + 2t)e^t \\ y = te^t. \end{cases}$$

This is our second solution.

Since it is clear from inspection that the two solutions we have found are linearly independent, we conclude that

$$\begin{cases} x = 2c_1 e^t + c_2(1 + 2t)e^t \\ y = c_1 e^t + c_2 t e^t \end{cases}$$

is the general solution of our system of differential equations.

☞ **You Try It:**   Find the general solution of the system

$$\begin{cases} \dfrac{dx}{dt} = 2x - 1y \\[2mm] \dfrac{dy}{dt} = x + 4y. \end{cases}$$

Math Note: We can think of a falling body as described by a linear system of differential equations. Let $x(t)$ be the height of the body at time t and let $y(t)$ be the velocity of the body at time t. Write down the system that describes a falling body that is dropped from height 50 ft (take the gravitational constant to be $g \approx -32$ ft/sec).

8.4 Nonlinear Systems: Volterra's Predator–Prey Equations

Imagine an island inhabited by foxes and rabbits. Foxes eat rabbits; rabbits, in turn develop methods of evasion to avoid being eaten. The resulting interaction is a fascinating topic for study, and is amenable to analysis via differential equations.

To appreciate the nature of the dynamic between the foxes and the rabbits, let us describe some of the vectors at play. We take it that the foxes eat rabbits—that is their source of food—and the rabbits eat exclusively clover. We assume that there is an endless supply of clover; the rabbits never run out of food. When the rabbits are abundant, then the foxes flourish and their population grows. When the foxes become too numerous and eat too many rabbits, then the rabbit population declines; as a result, the foxes enter a period of famine and their population begins to decline. As the foxes decrease in number, the rabbits become relatively safe and their population starts to increase again. This triggers a new increase in the fox population—as the foxes now have an increased source of food. As time goes on, we see an endlessly repeating cycle of interrelated increases and decreases in the populations of the two species. See Fig. 8.2, in which the sizes of the populations (x for rabbits, y for foxes) are plotted against time.

Problems of the sort that we have described here have been studied, for many years, by both mathematicians and biologists. It is pleasing to see how the mathematical analysis confirms the intuitive perception of the situation as described above. In our analysis below, we shall follow the approach of Vito Volterra (1860–1940), who was one of the pioneers in this subject.

If x is the number of rabbits at time t, then the relation

$$\frac{dx}{dt} = ax, \qquad a > 0$$

should hold, provided that the rabbits' food supply is unlimited and there are no foxes. This simply says that the rate of increase of the number of rabbits is proportional to the number present. [You studied equations of this kind in your calculus class when you learned about exponential growth.]

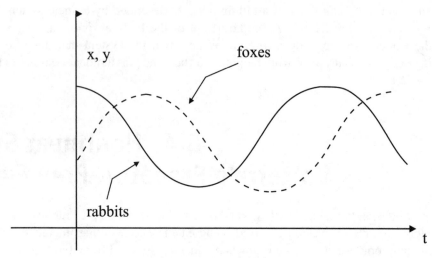

Fig. 8.2.

It is natural to assume that the number of "encounters" between rabbits and foxes per unit of time is jointly proportional to x and y. If we furthermore make the plausible assumption that a certain proportion of those encounters results in a rabbit being eaten, then we have

$$\frac{dx}{dt} = ax - bxy, \qquad a, b > 0.$$

In the same way, we notice that in the absence of rabbits the foxes die out, and their increase depends on the number of encounters with rabbits. Thus the same logic leads to the companion differential equation

$$\frac{dy}{dt} = -cy + gxy, \qquad c, g > 0.$$

We have derived the following nonlinear system describing the interaction of the foxes and the rabbits:

$$\begin{cases} \dfrac{dx}{dt} = x(a - by) \\[2mm] \dfrac{dy}{dt} = -y(c - gx). \end{cases} \qquad (1)$$

The equations (1) are called *Volterra's predator–prey equations*. It is a fact that this system cannot be solved explicitly in terms of elementary functions. On the other hand, we can perform what is known as a *phase plane analysis* and learn a great deal about the behavior of $x(t)$ and $y(t)$.

To be more specific, instead of endeavoring to describe x as a function of t and y as a function of t, we instead think of

$$\begin{cases} x = x(t) \\ y = y(t) \end{cases}$$

as the parametric equations of a curve in the x–y plane. We shall be able to determine the rectangular equations of this curve.

We begin by eliminating t in (1) and separating the variables. Thus

$$\frac{dx}{x(a - by)} = dt$$

$$\frac{dy}{-y(c - gx)} = dt,$$

hence

$$\frac{dx}{x(a - by)} = \frac{dy}{-y(c - gx)}$$

or

$$\frac{(a - by)\, dy}{y} = -\frac{(c - gx)\, dx}{x}.$$

Integration now yields

$$a \ln y - by = -c \ln x + gx + C.$$

In other words,

$$y^a e^{-by} = e^C x^{-c} e^{gx}. \tag{2}$$

If we take it that $x(0) = x_0$ and $y(0) = y_0$, then we may solve this last equation for e^C and find that

$$e^C = x_0^c y_0^a e^{-gx_0 - by_0}.$$

It is convenient to let $e^C = K$.

In fact we cannot solve (2) for either x or y. But we can utilize an ingenious method of Volterra to find points on the curve. To proceed, we give the left-hand side of (2) the name of z and the right-hand side the name of w. Then we plot the graphs C_1 and C_2 of the functions

$$z = y^a e^{-by} \qquad \text{and} \qquad w = K x^{-c} e^{gx} \tag{3}$$

as shown in Fig. 8.3. Since $z = w$ (by (2)), we must in the third quadrant depict this relationship with the dotted line L. To the maximum value of z given by the point

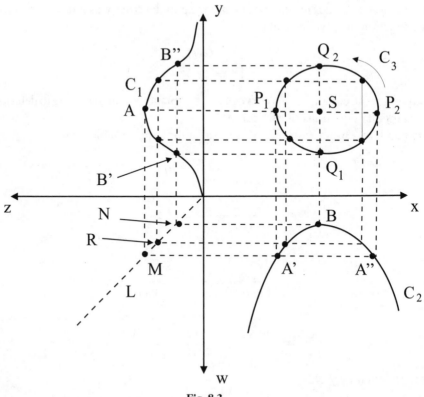

Fig. 8.3.

A on C_1, there corresponds one value of y and—via M on L and the corresponding points A' and A'' on C_2—two x's; and these determine the bounds between which x may vary.

Similarly, the minimum value of w given by B on C_2 leads to N on L and hence to B' and B'' on C_1; these points determine the limiting values for y. In this way we find the points P_1, P_2, and Q_1, Q_2 on the desired curve C_3. Additional points are easily found by starting on L at a point R (let us say) anywhere between M and N and projecting up to C_1 and over to C_3, and then over to C_2 and up to C_3. Again see Fig. 8.3.

It is clear that changing the value of K raises or lowers the point B, and this in turn expands or contracts the curve C_3. Accordingly, when K is given a range of values, then we obtain a family of ovals about the point S; and this is all there is of C_3 when the minimum value of w equals the maximum value of z.

We next show that, as t increases, the corresponding point (x, y) on C_3 moves around the curve in a counterclockwise direction. To see this, we begin by observing that equations (1) give the horizontal and vertical components of the velocity at this point. A simple calculation based on (3) shows that the point S has coordinates $x = c/g$, $y = a/b$. Namely, at those particular values of x and y, we see from (1) that both dx/dt and dy/dt are 0. Thus we must be at the stationary point S.

When $x < c/g$, the second equation of (1) tells us that dy/dt is negative, so that our point on C_3 moves down as it traverses the arc $Q_2 P_1 Q_1$. By similar reasoning, it moves up along the arc $Q_1 P_2 Q_2$. This proves our assertion.

We close this section by using the fox–rabbit system to illustrate the important method of *linearization*. First note that if the rabbit and fox populations are, respectively, constantly equal to

$$x = \frac{c}{g} \quad \text{and} \quad y = \frac{a}{b}, \tag{4}$$

then the system (1) is satisfied and we have $dx/dt \equiv 0$ and $dy/dt \equiv 0$. Thus there is no increase or decrease in either x or y. The populations (4) are called *equilibrium populations*; the populations x and y can maintain themselves indefinitely at these constant levels. This is the special case in which the minimum of w equals the maximum of z, so that the oval C_3 reduces to the point S.

We now return to the general case and put

$$x = \frac{c}{g} + X \quad \text{and} \quad y = \frac{a}{b} + Y;$$

here we think of X and Y as the deviations of x and y from their equilibrium values. An easy calculation shows that if we replace x and y in (1) with X and Y (which simply amounts to translating the point $(c/g, a/b)$ to the origin), then (1) becomes

$$\begin{cases} \dfrac{dX}{dt} = -\dfrac{bc}{g}Y - bXY \\[2mm] \dfrac{dY}{dt} = \dfrac{ag}{b}X + gXY. \end{cases} \tag{5}$$

The process of linearization now consists of assuming that if X and Y are small, then the XY term in (5) can be treated as negligible and hence discarded. This process results in (5) simplifying to the linear system (hence the name)

$$\begin{cases} \dfrac{dX}{dt} = -\dfrac{bc}{g}Y \\[2mm] \dfrac{dY}{dt} = \dfrac{ag}{b}X. \end{cases} \tag{6}$$

It is straightforward to solve (5) by the methods developed in this chapter. Easier still is to divide the left sides and right sides, thus eliminating dt, to obtain

$$\frac{dY}{dX} = -\frac{ag^2}{b^2c}\frac{X}{Y}.$$

The solution of this last equation is immediately seen to be

$$ag^2 X^2 + b^2 c Y^2 = C^2.$$

This is a family of ellipses centered at the origin in the X–Y plane. Since ellipses are qualitatively similar to the ovals of Fig. 8.3, we may hope that (6) is a reasonable approximation to (5).

 Math Note:　Of course foxes and rabbits are a simple-minded paradigm for predator–prey systems. We could instead use the ideas presented here to study competing software companies, or professors competing for grants, or forest fires and forests. The ideas initiated by Volterra one hundred years ago have become an established and prominent part of mathematical analysis of real-world situations.

One of the important themes that we have introduced in this chapter, which arose naturally in our study of systems, is that of nonlinearity. Nonlinear equations have none of the simple structure, nor any concept of "general solution," that the more familiar linear equations have. They are currently a matter of intense study.

In studying a system like (1), we have learned to direct our attention to the behavior of solutions near points in the x–y plane at which the right sides both vanish. We have seen why periodic solutions (i.e., those that yield simple closed curves like C_3 in Fig. 8.3) are important and advantageous for our analysis. And we have given a brief hint of how it can be useful to study a nonlinear system by approximation with a linear system.

Exercises

1.　Replace each of the following differential equations by an equivalent system of first-order equations:

(a) $y'' - xy' - xy = 0$

(b) $y''' = y'' - x^2(y')^2$

(c) $xy'' - x^2 y' - x^3 y = 0$

2. (a) Show that

$$\begin{cases} x = e^{4t} \\ y = e^{4t} \end{cases} \quad \text{and} \quad \begin{cases} x = e^{-2t} \\ y = -e^{-2t} \end{cases}$$

are solutions of the homogeneous system

$$\begin{cases} \dfrac{dx}{dt} = x + 3y \\ \dfrac{dy}{dt} = 3x + y. \end{cases}$$

(b) Find the particular solution

$$\begin{cases} x = x(t) \\ y = y(t) \end{cases}$$

of this system for which $x(0) = 5$ and $y(0) = 1$.

3. (a) Show that

$$\begin{cases} x = 2e^{4t} \\ y = 3e^{4t} \end{cases} \quad \text{and} \quad \begin{cases} x = e^{-t} \\ y = -e^{-t} \end{cases}$$

are solutions of the homogeneous system

$$\begin{cases} \dfrac{dx}{dt} = x + 2y \\ \dfrac{dy}{dt} = 3x + 2y. \end{cases}$$

(b) Show that

$$\begin{cases} x = 3t - 2 \\ y = -2t + 3 \end{cases}$$

is a particular solution of the nonhomogeneous system

$$\begin{cases} \dfrac{dx}{dt} = x + 2y + t - 1 \\ \dfrac{dy}{dt} = 3x + 2y - 5t - 2. \end{cases}$$

Write the general solution of this system.

4. Use the methods treated in this chapter to find the general solution of each of the following systems:

    (a) $\begin{cases} \dfrac{dx}{dt} = -3x + 4y \\[2mm] \dfrac{dy}{dt} = -2x + 3y \end{cases}$

    (b) $\begin{cases} \dfrac{dx}{dt} = 2x \\[2mm] \dfrac{dy}{dt} = 3y \end{cases}$

5. Replace each of the following ordinary differential equations by an equivalent system of first-order equations:

    (a) $y''' + x^2 y'' - xy' + y = x$

    (b) $y'' - [\sin x]y' + [\cos x]y = 0$

6. In each of the following problems, show that the given solution set indeed satisfies the system of differential equations:

    (a) $x(t) = \frac{1}{2}Ae^{-5t} + 2Be^{t}, \qquad y(t) = Ae^{-5t} + Be^{t}$

    $x'(t) = 3x(t) - 4y(t)$

    $y'(t) = 4x(t) - 7y(t)$

    (b) $x(t) = Ae^{3t} + Be^{-t}, \qquad y(t) = 2Ae^{3t} - 2Be^{-t}$

    $x'(t) = x(t) + y(t)$

    $y'(t) = 4x(t) + y(t)$

7. Solve each of the following systems of linear ordinary differential equations:

    (a) $x'(t) = 3x(t) + 2y(t)$

    $y'(t) = -2x(t) - y(t)$

    (b) $x'(t) = x(t) + y(t)$

    $y'(t) = -x(t) + y(t)$

8. Solve the initial value problem

    $$\begin{cases} x' = y \\ y' = x \end{cases}$$

    with $x(0) = 1,\ y(1) = 0$.

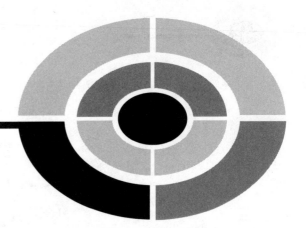

Final Exam

1. The general solution of the differential equation $y'' - 4y = 0$ is
   (a) $y = Ae^{2x} + Bd^{-2x}$
   (b) $y = e^{2x}, \ y = e^{-2x}$
   (c) $y = e^{2x} + e^{-2x}$
   (d) $y = Ae^{x} + Be^{-x}$
   (e) $y = Ae^{x} + Bd^{2x}$

2. A solution of the differential equation $(y')^2 x - 9x^2 y = 0$ is given by
   (a) $y = x^2$
   (b) $y = x + 1$
   (c) $y = x^3$
   (d) $y = x^2 - x$
   (e) $y = \cos x$

3. The solution of the initial value problem $y' + xy = x$, $y(0) = 2$, is given by
   (a) $y = e^{x^2/2}$
   (b) $y = xe^{-x}$
   (c) $y = x + Ce^{-x^2}$
   (d) $y = 1 + e^{-x^2/2}$
   (e) $y = x^2 - e^{-x^2}$

4. A solution of the differential equation $y' \cdot \sin y = x \ln x$ is given by

(a) $y = \arccos\left[-\dfrac{x^2}{2}\ln x + \dfrac{x^2}{4} + C\right]$

(b) $y = \arcsin\left[\dfrac{x^3}{6} - \ln x\right]$

(c) $y = \arctan\left[\dfrac{\ln x}{x} - x^3\right]$

(d) $y = \arccos\left[\dfrac{\ln x}{x^2} - x^2 \ln x\right]$

(e) $y = \arcsin\left[x\ln x + \dfrac{x^2}{\ln x}\right]$

5. Of the differential equations (i) $y'' - (\sin x)y = e^x$, (ii) $y^{(iv)} + x^2 y'' - e^x y = \cos x$, (iii) $y''' + y'' + x^5 y = x^3$, and (iv) $x^4 y'' + x^3 y'' + x^2 y = x^5$, we see that

(a) Equation (i) is of second order, equation (ii) is of third order, equation (iii) is of fifth order, and equation (iv) if of fourth order.

(b) Equation (i) is of second order, equation (ii) is of fourth order, equation (iii) is of third order, and equation (iv) is of second order.

(c) Equation (i) is of third order, equation (ii) is of fourth order, equation (iii) is of second order, and equation (iv) if of first order.

(d) Equation (i) is of fourth order, equation (ii) is of third order, equation (iii) is of second order, and equation (iv) is of first order.

(e) Equation (i) is of first order, equation (ii) is of second order, equation (iii) is of third order, and equation (iv) is of fourth order.

6. Use the method of separation of variables to completely solve the differential equation $xy' = \ln x \cdot y$.

(a) $y = C \cdot x e^{x^2}$

(b) $y = C \cdot x^2 \ln x$

(c) $y = C \cdot x / \ln x$

(d) $y = C \cdot \ln x / x^2$

(e) $y = C \cdot e^{\ln^2 x/2}$

7. Use the method of separation of variables to solve the initial value problem $xy' = y^2/x^2$, $y(1) = 4$.

(a) $y = C/[\ln x + 1]$

(b) $y = 1/[x^{-2}/2 - 1/4]$

(c) $y = x/[x^{-1} - x^2]$

(d) $y = x/[x + 1]$

(e) $y = x \ln^2 x$

8. Use the method of first-order linear equations to find the general solution of the equation $y' - y/x = 1$.
   (a) $y = Ce^x + x$
   (b) $y = Cx^2 + x$
   (c) $y = x - C \ln x$
   (d) $y = x \ln x + Cx$
   (e) $y = C \ln x - x$

9. Use the method of first-order linear equations to find the unique solution to the initial value problem $y' - (\cos x)y = \cos x$ $y(0) = 1$.
   (a) $y = (\cos x)e^{\sin x}$
   (b) $y = (\sin x)e^{-\cos x}$
   (c) $y = x \cos x$
   (d) $y = (\cos x)(\sin x)$
   (e) $y = -1 + 2e^{\sin x}$

10. The differential equation $x^2 y'' - \sin x(y')^2 + (\ln x)y = e^x$ is not linear because
    (a) there is an x^2 factor in front of the lead term.
    (b) there is a factor of $\sin x$ in front of y'.
    (c) the term y' is squared.
    (d) there is a factor of $\ln x$ in front of y.
    (e) there is a term e^x on the right.

11. Of the differential equations (i) $xy^2 \, dx - yx^2 \, dy = 0$, (ii) $x \cos y \, dx + y \sin x \, dy = 0$, (iii) $[y^2 \cos xy - xy^2 \sin xy] \, dx + [2xy \cos xy - x^2y^2 \sin xy] \, dy = 0$, and (iv) $3x^2 y \, dx + x^3 \, dy = 0$, we see that
    (a) Equations (i), (iii), (iv) are exact.
    (b) Equations (ii), (iii), (iv) are exact.
    (c) Equations (iii) and (iv) are exact.
    (d) Equations (i) and (ii) are exact.
    (e) Equations (ii) and (iii) are exact.

12. Use the method of exact equations to solve the differential equation $2xy^3 \, dx + 3y^2 x^2 \, dy = 0$.
    (a) $y = C \cdot x^{2/3}$
    (b) $y = C \cdot x^2$
    (c) $y = C \cdot x^{-2/3}$
    (d) $y = C \cdot x^{1/3}$
    (e) $y = C \cdot x^{-1/3}$

13. Find the orthogonal trajectories to the family of curves $y = Cx^3$.
    (a) $y = \sqrt{-\dfrac{x^2}{3} + E}$
    (b) $y = -1/x + D$

(c) $y = x^2 + E$

(d) $y = Cx + D$

(e) $y^2 + x^2 = C$

14. Of the differential equations (i) $xy - y^2 = dy/dx$, (ii) $(x^2 - xy) dx - (y^2 + xy) dy = 0$, (iii) $(x/y - \sin(x/y)) dx + (y^2/x^2 + \ln(x/y)) dy = 0$, and (iv) $x\,dy - y\,dx = 0$, which are homogeneous?

    (a) Equations (i), (ii), and (iii).

    (b) Equations (i), (iii), and (iv).

    (c) Equations (i) and (iv).

    (d) Equations (iii) and (iv).

    (e) Equations (ii), (iii), and (iv).

15. Find an integrating factor for the differential equation $[2y/x]\,dx + 1\,dy = 0$ and then solve the equation. The solution is

    (a) $y = Cx^2$

    (b) $y = Cx + x^2$

    (c) $y = C/x^2$

    (d) $y = C + x^2$

    (e) $y = Cx - x^2$

16. Use the method of reduction of order to solve the differential equation $y'' + y' = x$.

    (a) $y = x^2 + x - e^x$

    (b) $y = x^2 - x + Ce^{-x}$

    (c) $y = x + Ce^x$

    (d) $y = x^2/2 - x + Ce^{-x}$

    (e) $y = e^x - Cx^2$

17. Use the method of reduction of order to find some solution to the equation $y'' - y^2 = 0$.

    (a) $y = 1/x - 1/x^2$

    (b) $y = x + x^2$

    (c) $y = x - 1/x$

    (d) $y = x + 1/x^2$

    (e) $y = 6/x^2$

18. The method of reduction of order will not work on the equation $(\sin x)y' - x^2 y = e^x$ because

    (a) there is a nonlinear term on the right-hand side.

    (b) there is a factor of x^2 in front of the y.

    (c) the lead coefficient is not 1.

    (d) the equation is not of second order.

    (e) the equation is too complex.

19. The general solution of the differential equation $y'' - 5y' + 4y = 0$ is given by
   (a) $y = Ae^x + Be^{-x}$
   (b) $y = Ae^{4x} + Be^x$
   (c) $y = Ae^{-x} + B^{-2x}$
   (d) $y = Ae^x + Be^{2x}$
   (e) $y = Ae^{4x} + Be^{-4x}$

20. The solution of the initial value problem $y'' + 5y' + 6y = 0$, $y(0) = 1$, $y'(0) = 2$, is given by
   (a) $y = 5e^{-2x} - 4e^{-3x}$
   (b) $y = 3e^{-2x} + 5e^{-3x}$
   (c) $y = 4e^{2x} + 3e^{3x}$
   (d) $y = e^{-2x} + e^{-3x}$
   (e) $y = -2e^{-2x} - 3e^{-3x}$

21. Two linearly independent solutions of the differential equation $y'' + 2y' + 10y = 0$ are given by
   (a) $y_1 = e^{-x} \sin 3x$, $y_2 = e^{-x} \cos 3x$
   (b) $y_1 = e^x \sin 6x$, $y_2 = e^x \sin 4x$
   (c) $y_1 = e^x \sin 6x$, $y_2 = e^x \cos 4x$
   (d) $y_1 = e^{-x} \sin 4x$, $y_2 = e^{-x} \cos 4x$
   (e) $y_1 = \sin x$, $y_2 = \cos x$

22. The general solution of the differential equation $y'' - 7y' + 10y = x$ is
   (a) $y = Ae^{2x} + Be^{5x}$
   (b) $y = Ae^{2x} + Be^{5x} + x^2 - x$
   (c) $y = Ae^{-2x} + Be^{-5x} + (x/10 - 7/10)$
   (d) $y = Ae^{2x} + Be^{5x} + (x/10 - 7/10)$
   (e) $y = A \cos 2x + B \sin 5x + x$

23. The unique solution of the initial value problem $y'' - 5y' + 6y = x^2$, $y(0) = 2$, $y'(0) = 1$, is
   (a) $2e^{2x} - 3e^{3x} + \dfrac{x^2}{6} - x$
   (b) $e^{2x} - e^{3x} + x^2 + x$
   (c) $\dfrac{513}{108}e^{2x} - \dfrac{79}{27}e^{3x} + \dfrac{x^2}{6} + \dfrac{5x}{18} + \dfrac{19}{108}$
   (d) $\dfrac{11}{108}e^{2x} - \dfrac{29}{27}e^{3x} + \dfrac{x^2}{2} - \dfrac{x}{6}$
   (e) $12e^x - 6e^{-x} + x^2 - x + 4$

24. A solution of the differential equation $x^2 y'' - xy' + y = x^2$ is given by
    (a) $y = 3x^2 - x$
    (b) $y = x^2 + x$
    (c) $y = -x^2 + x^3$
    (d) $y = x - x^4$
    (e) $y = 4 + x^2$

25. The general solution of the differential equation $y'' - 4y' + 4y = 0$ is
    (a) $y = Ae^{2x} + Be^{3x}$
    (b) $y = Ae^{2x} + Bxe^{2x}$
    (c) $y = Ae^{2x} + B\sin 2x$
    (d) $y = A\sin 2x + B\cos 2x$
    (e) $y = Ae^{2x} + Be^{-2x}$

26. Use the method of variation of parameters to find the general solution of the differential equation $y'' - 7y' + 12y = e^x$.
    (a) $y = Ae^{3x} + Be^{4x} + e^x$
    (b) $y = Ae^x + Be^{3x} + e^x/6$
    (c) $y = Ae^{3x} + Be^{-3x} + e^{4x}$
    (d) $y = Ae^{3x} + Be^{4x} + e^x/6$
    (e) $y = Ae^{-3x} + Be^{-4x} + e^x/6$

27. Use the method of undetermined coefficients to find the general solution of the differential equation $y'' - 7y' + 12y = \cos x$.
    (a) $y = Ae^{3x} + Be^{4x} + \cos x$
    (b) $y = Ae^{3x} + Be^{4x} + \sin x$
    (c) $y = Ae^{-3x} + Be^{-4x} + \cos x$
    (d) $y = Ae^{-3x} + Be^{-4x} + \sin x$
    (e) $y = Ae^{3x} + Be^{4x} + \dfrac{11\cos x}{170} - \dfrac{7\sin x}{170}$

28. Find the general solution of the differential equation $y''' - 2y'' - 5y' + 6$.
    (a) $y = Ae^x + Be^{2x} + Ce^{3x}$
    (b) $y = Ae^x + Be^{-2x} + Ce^{3x}$
    (c) $y = Ae^x + Be^{-2x}$
    (d) $y = Ae^{-2x} + Be^{3x}$
    (e) $y = Ae^{2x} + Be^{-x} + Ce^x$

29. Given that the differential equation $y'' - (1/x)y' = 0$ has the function $y \equiv 1$ as a solution, find the general solution.
    (a) $y = Ax^2 + B$
    (b) $y = Ax^2 + Bx + C$

(c) $y = Ax^2 + Cx$

(d) $y = Ax + B$

(e) $y = A + B$

30. Kepler's Second Law tells us that, the more eccentric the elliptical orbit of a planet,

    (a) the greater the speed of the planet when it traverses the portion of the orbit that is flattest (i.e., has least curvature).

    (b) the lesser the speed of the planet when it traverses the portion of the orbit that is flattest (i.e., has least curvature).

    (c) the more likely the planet is to "jump orbit" and float off into space.

    (d) the more likely the planet is to slow down at the end of the day.

    (e) the more likely the planet is to speed up when it passes the earth.

31. Kepler's Third Law tells us that the length of a year on Venus is

    (a) longer than an Earth year.

    (b) the same length as an Earth year.

    (c) shorter than an Earth year.

    (d) elliptical in shape.

    (e) variable.

32. The radius of convergence of the power series $\sum_{j=0}^{\infty}[2^j x^j]/j!$ is

    (a) 4

    (b) $+\infty$

    (c) 1

    (d) 0

    (e) 2

33. Calculate the power series expansion of the function $f(x) = x \cdot \cos x^2$ about the point $c = 0$.

    (a) $f(x) = \sum_{j=0}^{\infty} \dfrac{x^j}{j^2}$

    (b) $f(x) = \sum_{j=0}^{\infty} 3^{-j} x^{2j}$

    (c) $f(x) = \sum_{j=0}^{\infty} (-1)^j \cdot \dfrac{x^{4j+1}}{(2j)!}$

    (d) $f(x) = \sum_{j=0}^{\infty} (-1)^j \cdot \dfrac{x^{2j+1}}{(2j)!}$

    (e) $f(x) = \sum_{j=0}^{\infty} \dfrac{x^{2j}}{j!}$

34. According to the method of power series, the solution of the differential equation $y' - xy = x$ is

   (a) $y = \sum_{j=0}^{\infty} \dfrac{x^j}{(2j)!}$

   (b) $y = \sum_{j=0}^{\infty} \dfrac{x^j}{j}$

   (c) $y = \sum_{j=1}^{\infty} \dfrac{x^{2j+1}}{(2j+1)!}$

   (d) $y = \sum_{j=1}^{\infty} \dfrac{x^{3j}}{(2j)!}$

   (e) $y = \dfrac{1}{2} + \sum_{j=1}^{\infty} \dfrac{x^{2j}}{j!}$

35. According to the method of power series, the solution of the differential equation $y'' + y = x^2$ is

   (a) $y = x^2 - 2 + A \sum_{j=0}^{\infty}(-1)^j \dfrac{x^{2j}}{(2j)!} + B \sum_{j=0}^{\infty}(-1)^j \dfrac{x^{2j+1}}{(2j+1)!}$

   (b) $y = x + 1 + A \sum_{j=0}^{\infty}(-1)^j \dfrac{x^{2j}}{(2j)!} + B \sum_{j=0}^{\infty}(-1)^j \dfrac{x^{2j+1}}{(2j+1)!}$

   (c) $y = x^2 - 2x + A \sum_{j=0}^{\infty} \dfrac{x^j}{j} + B \sum_{j=0}^{\infty} \dfrac{x^{2j}}{j!}$

   (d) $y = x + x^3 - A \sum_{j=0}^{\infty} \dfrac{2^j}{j!} x^j$

   (e) $y = x^2 + x - 1 + A \sum_{j=0}^{\infty} \dfrac{2^j}{(j+3)!} x^j + B \sum_{j=0}^{\infty} \dfrac{3^{-j}}{j!} x^{2j+1}$

36. The method of power series tells us that the general solution of the differential equation $y' = y$ is
   (a) $y = Ce^{3x}$
   (b) $y = Ce^{-x}$
   (c) $y = C \sin x$
   (d) $y = C \cos x$
   (e) $y = Ce^x$

37. The recursion relations for the coefficients of the power series solution to the differential equation $y'' - xy = x$ are

   (a) $a_0 = 0; a_{j+2} = \dfrac{1}{j(j-1)} a_j, j \geq 1$

   (b) $a_{j+2} = \dfrac{j}{(j+2)j}, j \geq 0$

   (c) $a_{j+1} = j \cdot a_j, j \geq 1$

(d) $a_2 = 0, a_3 = (1 + a_0)/(3 \cdot 2), a_{j+3} = a_j/((j+3)(j+2)), j \geq 1$

(e) $a_2 = a_0 + 2a_1, a_{j+2} = [a_j - a_{j-1}]/6$

38. With the method of power series, the solution to the initial value problem $y' + xy = x, y(0) = 3$ is

   (a) $y = \sum_{j=0}^{\infty} \dfrac{x^{2j}}{(j+1)!}$

   (b) $y = 1 + 2\sum_{j=0}^{\infty}(-1)^j \dfrac{x^{2j}}{j!}$

   (c) $y = \sum_{j=0}^{\infty} \dfrac{3x^j}{e^j}$

   (d) $y = -3 + \sum_{j=0}^{\infty} \dfrac{x^{2j+1}}{(j+2)!}$

   (e) $y = \dfrac{x}{2} - \sum_{j=0}^{\infty}(-1)^{j+1}\dfrac{j! \cdot x^j}{j^j}$

39. Begin with the geometric series in x and find a power series representation for the function $1/(1-x)^3$.

   (a) $\sum_{j=0}^{\infty} j(j+1)x^{j-1}$

   (b) $\sum_{j=0}^{\infty} \dfrac{j!}{j^j}x^j$

   (c) $\frac{1}{2}\sum_{j=0}^{\infty}(j+1)(j+2)x^j$

   (d) $\sum_{j=0}^{\infty} \dfrac{j}{j+1}x^{2j}$

   (e) $\sum_{j=0}^{\infty} j^2 x^{j-1}$

40. The coefficient of x^3 in the power series expansion solution of the initial value problem $y'' + xy' + y = 1, y(0) = 2, y'(0) = 1$, is

   (a) $-1/3$

   (b) $1/3$

   (c) $2/5$

   (d) $1/7$

   (e) $2/9$

41. The Fourier series of the function $f(x) = x^2$ on the interval $[-\pi, \pi)$ is

   (a) $\pi^2 + \sum_{j=1}^{\infty} \dfrac{j\pi}{2}\cos jx$

   (b) $\dfrac{\pi^2}{3} + \sum_{j=1}^{\infty} \dfrac{(-1)^j 4}{j^2}$

   (c) $\sum_{j=1}^{\infty} \pi^3 \sin jx + \sum_{j=1}^{\infty} \pi^2 \cos jx$

(d) $\dfrac{\pi}{6} + \sum_{j=1}^{\infty} \cos jx - \sin jx$

(e) $\dfrac{-2\pi}{5} + \sum_{j=1}^{\infty} j^2 \cos jx + j \sin jx$

42. The Fourier series of the function $f(x) = \cos^2 x - 3 \sin^2 x$ on the interval $[-\pi, \pi]$ is

(a) $1 + \sin 2x$

(b) $3 - \cos 4x$

(c) $2 + \sin 3x$

(d) $2 - \cos 2x$

(e) $4 + 2 \cos 2x$

43. The Fourier series of the function $f(x) = x$ on the interval $[-2, 2]$ is

(a) $\sum_{j=1}^{\infty} (j^2 - j) \sin \dfrac{j\pi x}{2}$

(b) $\sum_{j=1}^{\infty} (j^2 + j) \sin \dfrac{j\pi x}{2}$

(c) $\sum_{j=1}^{\infty} \dfrac{j^3}{2} \sin \dfrac{j\pi x}{2}$

(d) $\sum_{j=1}^{\infty} j \sin(2jx)$

(e) $\sum_{j=1}^{\infty} \dfrac{-4(-1)^j}{j\pi} \sin \dfrac{j\pi x}{2}$

44. The function

$$f(x) = \begin{cases} 1 & \text{if } -3 \le x < 0 \\ -1 & \text{if } 0 \le x \le 3 \end{cases}$$

has Fourier series with only sine terms (no cosine terms appear). This is so because

(a) The function f is locally constant.

(b) The function f is bounded by 2.

(c) The function f is not periodic.

(d) The function f is odd.

(e) The function f is piecewise linear.

45. The functions $f(x) = x$ and $g(x) = x^2$ are orthogonal on the interval $[-1, 1]$ in the sense that

$$\int_{-1}^{1} f(x)g(x)\,dx = 0.$$

Find a third function h, a third-degree polynomial, that is orthogonal to both f and g.

(a) $h(x) = x^3 - 2x^2 + x$

(b) $h(x) = \dfrac{x^3}{4} - \dfrac{x^2}{3} + 2x$

(c) $h(x) = \dfrac{2x^3}{3} + \dfrac{2x^2}{3} - \dfrac{2x}{5} - \dfrac{2}{5}$

(d) $h(x) = \dfrac{x^3}{2} + \dfrac{2x^2}{7} - x$

(e) $h(x) = x^3 + 5x^2 = 3x + 2$

46. Find the cosine series for the function $f(x) = \sin 2x$ on the interval $[0, \pi]$.

(a) $f(x) = \sum_{j=1}^{\infty} \dfrac{j}{j^2 + 4} \cos jx$

(b) $f(x) = \sum_{j=1}^{\infty} f(j + 1) \cos jx$

(c) $f(x) = \sum_{j=1}^{\infty} \dfrac{4}{\pi} \left[\dfrac{1 + (-1)^{j+1}}{4 - j^2} \right] \cos jx$

(d) $f(x) = \sum_{j=1}^{\infty} \dfrac{j}{\pi} (-1)^{j+1} \cos jx$

(e) $f(x) = \sum_{j=1}^{\infty} \dfrac{j^2 + j}{4} \cos jx$

47. Find the Fourier series of the function

$$f(x) = \begin{cases} 0 & \text{if } -\pi \le x < 0 \\ 1 & \text{if } 0 \le x \le \pi/2 \\ 0 & \text{if } \pi/2 < x \le \pi. \end{cases}$$

(a) $\sum_{j=1}^{\infty} \dfrac{j^2}{2} \cos jx + \sum_{j=1}^{\infty} \dfrac{j}{2} \sin jx$

(b) $\sum_{j=1}^{\infty} \dfrac{\pi j}{j + 1} \cos jx + \sum_{j=1}^{\infty} \dfrac{\pi}{2j} \sin jx$

(c) $\dfrac{1}{4} + \sum_{j=1}^{\infty} \dfrac{1}{2j} (-1)^{[j/2]} \left[(-1)^{j+1} + 1 \right] \cos jx$

$+ \sum_{j=1}^{\infty} \dfrac{-1}{2j} (-1)^{[j/2]} \left[(-1)^{j} + 1 \right] \sin jx$

(d) $\sum_{j=1}^{\infty} j \cos jx + \sum_{j=1}^{\infty} (j + 1) \sin jx$

(e) $\sum_{j=1}^{\infty} \dfrac{1}{j} \cos jx + \sum_{j=1}^{\infty} \dfrac{1}{j + 1} \sin jx$

48. Find the Fourier series of the function $f(x) = e^x$ on the interval $[-\pi, \pi]$.

(a) $\dfrac{1}{2\pi} e^{\pi} + \sum_{j=1}^{\infty} \dfrac{j}{j^2 + 1} \cos jx$

$+ \sum_{j=1}^{\infty} \dfrac{1}{j^2 + 1} \sin jx$

(b) $\frac{1}{2\pi}\left(e^{\pi}-e^{-\pi}\right)+\sum_{j=1}^{\infty}\frac{(-1)^{j}e^{\pi}}{\pi}\cdot\frac{1}{1+j^{2}}\cos jx$

$+\sum_{j=1}^{\infty}\frac{(-1)^{j}e^{-\pi}}{\pi}\cdot\frac{j}{1+j^{2}}\sin jx$

(c) $\frac{1}{2\pi}\left(e^{-\pi}-e^{\pi}\right)+\sum_{j=1}^{\infty}\frac{(-1)^{j}e^{\pi}}{\pi}\cdot\frac{j}{1+j^{2}}\cos jx$

$+\sum_{j=1}^{\infty}\frac{(-1)^{j}e^{-\pi}}{\pi}\cdot\frac{1}{1+j^{2}}\sin jx$

(d) $\frac{1}{2\pi}\left(e^{\pi/2}-e^{-\pi/2}\right)+\sum_{j=1}^{\infty}\frac{(-1)^{j+1}e^{2\pi}}{\pi}\cdot\frac{1}{1+j^{2}}\cos jx$

$+\sum_{j=1}^{\infty}\frac{(-1)^{j+1}e^{-2\pi}}{2\pi}\cdot\frac{j}{1+j^{2}}\sin jx$

(e) $\frac{1}{2\pi}\left(e^{3\pi}-e^{-3\pi}\right)+\sum_{j=1}^{\infty}\frac{(-1)^{j^{2}+1}e^{-\pi}}{\pi}\cdot\frac{j^{2}}{1+j^{2}}\cos jx$

$+\sum_{j=1}^{\infty}\frac{(-1)^{j^{2}+1}e^{\pi}}{2\pi}\cdot\frac{-j^{2}}{1+j^{2}}\sin jx$

49. Find the Fourier series of the function $f(x)=\sin x-4\cos 7x$.
    (a) $\sin x-4\cos 7x$
    (b) $\cos x-4\sin 7x$
    (c) $\cos 4x-\sin 7x$
    (d) $\sin 4x-\cos 7x$
    (e) $\sin 7x-\cos 4x$

50. Find the Fourier series of the function

$$f(x)=\begin{cases}0 & \text{if } -2\le x<0\\ 1 & \text{if } 0\le x\le 2\end{cases}$$

on the interval $[-2,2]$.

(a) $\sum_{j=1}^{\infty}\frac{j}{2}\cos jx+\sum_{j=1}^{\infty}\frac{2}{j}\sin jx$

(b) $\sum_{j=1}^{\infty}j^{2}\cos jx+\sum_{j=1}^{\infty}j\sin jx$

(c) $\sum_{j=1}^{\infty}j!\cos jx+\sum_{j=1}^{\infty}j^{j}\sin jx$

(d) $\sum_{j=1}^{\infty}\frac{-1}{j\pi}[(-1)^{j}-1]\sin\frac{j\pi x}{2}$

(e) $\sum_{j=1}^{\infty}\frac{1}{j\pi}[(-1)^{j}+1]\cos\frac{j\pi x}{2}$

51. Find the Fourier series of the function

$$f(x)=\begin{cases}0 & \text{if } x\le 0\\ x & \text{if } x>0\end{cases}$$

on the interval $[-\pi, \pi]$.

(a) $\dfrac{\pi}{4} + \sum_{j=1}^{\infty} \left[\dfrac{(-1)^j + 1}{j^2} \right] \cos jx + + \sum_{j=1}^{\infty} \left[\dfrac{\pi(-1)^{j+1}}{j} \right] \sin jx$

(b) $\sum_{j=1}^{\infty} (-1)^j \cdot j^2 \cdot \sin jx$

(c) $\sum_{j=1}^{\infty} (-1)^{j+1} \cdot j^3 \cos jx$

(d) $\sum_{j=1}^{\infty} \dfrac{j}{3} \cos jx + \sum_{j=1}^{\infty} \dfrac{j^2}{5} \sin jx$

(e) $\sum_{j=1}^{\infty} j(j+1) \cos jx$

52. The Fourier series, calculated as usual using the Riemann integral of calculus, of the function

$$ f(x) = \begin{cases} 0 & \text{if } x \text{ is rational} \\ 1 & \text{if } x \text{ is irrational} \end{cases} $$

(a) exists and is identically 0

(b) has only cosine terms

(c) has only sine terms

(d) has both sine and cosine terms, but only with even frequencies

(e) does not exist because the function f is not integrable

53. Find the eigenvalues λ_n and the eigenfunctions y_n for the equation $y'' + \lambda y = 0$ with the boundary conditions $y(0) = 0$, $y(\pi/3) = 0$.

(a) eigenvalues are $2, 4, 6, 8, \ldots$ and eigenfunctions are $\sin 2x, \sin 4x, \sin 6x, \sin 8x, \ldots$

(b) eigenvalues are $9, 36, 81, \ldots$ and eigenfunctions are $\sin 3x, \sin 6x, \sin 9x, \ldots$

(c) eigenvalues are $3, 6, 9, \ldots$ and eigenfunctions are $\sin 3x, \sin 6x, \sin 9x, \ldots$

(d) eigenvalues are $4, 9, 16, \ldots$ and eigenfunctions are $\sin 2x, \sin 3x, \sin 4x, \ldots$

(e) eigenvalues are $6, 9, 12, \ldots$ and eigenfunctions are $\sin \sqrt{6}x, \sin \sqrt{9}x, \sin \sqrt{12}x, \ldots$

54. Consider the wave equation $u_{tt} = u_{xx}$ with $a = 1$. Assume that the string has an initial configuration (before it is released to vibrate) given by $\varphi(x) = x^2 - \pi x$. Also the initial velocity is identically $\psi(x) = 0$. Then d'Alembert's solution to the equation is

(a) $u(x, t) = t^2 - x^2 + x$

(b) $u(x, t) = xt^2 - tx^2 + t^3$

(c) $u(x, t) = x^2 + t^2 - \pi x$

(d) $u(x, t) = xt - x + t$

(e) $u(x, t) = xt - x + t$

55. Solve the Dirichlet problem on the unit disc with boundary data $f(\theta) = \sin(\theta/2), -\pi \le \theta \le \pi$.

(a) $w(r, \theta) = \frac{1}{2} + \sum_{j=0}^{\infty} \frac{r^j}{j!} \cos j\theta$

(b) $w(r, \theta) = \sum_{j=1}^{\infty} r^j \frac{(-1)^j 8j}{\pi(1 - 4j^2)} \sin j\theta$

(c) $w(r, \theta) = \sum_{j=1}^{\infty} r^j \frac{(-1)^j 2^j}{j!} \cos j\theta + \sum_{j=1}^{\infty} r^j \frac{2^{-j}}{j!} \sin j\theta$

(d) $w(r, \theta) = \sum_{j=1}^{\infty} r^j j^3 \cos j\theta$

(e) $w(r, \theta) = \sum_{j=1}^{\infty} r^j \frac{j}{j+1} \sin j\theta$

56. Solve the Dirichlet problem on the unit disc with boundary data $f(\theta) = \cos^2 \theta$.

(a) $\frac{1}{2} + \frac{r^2}{2} \cos 2\theta$

(b) $\frac{1}{2} + r \sin \theta - r^2 \cos \theta$

(c) $\frac{1}{2} - \frac{r^2}{2} \cos 2\theta$

(d) $\frac{1}{2} + \frac{r^2}{2} \sin \theta$

(e) $\frac{1}{2} + \frac{r^3}{3} \cos 2\theta$

57. Consider the wave equation $u_{tt} = u_{xx}$ with $a = 1$. Assume that the string has an initial configuration $\varphi(x) = \sin x$ and an initial velocity $\psi(x) = 2x$. Then d'Alembert's solution to the equation is

(a) $u(x, t) = tx - \sin t \cos x$

(b) $u(x, t) = t^2 x + \cos t \sin x$

(c) $u(x, t) = 2tx + \cos t \sin x$

(d) $u(x, t) = tx^2 - \cos^2 t \sin x$

(e) $u(x, t) = t \sin x + x \cos t$

58. Find the eigenvalues and eigenfunctions for the equation $y'' + \lambda y = 0$ in the case $y(1) = 0, y(5) = 0$.

(a) eigenvalues are $1, 4, 9, \ldots$ and eigenfunctions are $\cos \pi x$, $\cos 2\pi x$, $\cos 3\pi x$, etc.

(b) eigenvalues are $2, 4, 6, 8, \ldots$ and eigenfunctions are $\cos 4x$, $\cos 8x$, $\cos 12x$, ...

(c) eigenvalues are $3, 6, 9, \ldots$ and eigenfunctions are $\sin 3x$, $\sin 6x$, $\sin 9x$, ...

(d) eigenvalues are $j^2\pi^2/16$ and eigenfunctions are $\cos\dfrac{\pi x}{4} - \sin\dfrac{\pi x}{4}$, $\cos\dfrac{\pi x}{2}$,
$\cos\dfrac{3\pi x}{4} + \sin\dfrac{3\pi x}{4}$, $\sin\pi x$, etc.

(e) eigenvalues are $4, 9, 16, \ldots$ and eigenfunctions are $\sin 2x$, $\sin 3x$, $\sin 4x, \ldots$.

59. The rod in our model for heat equation has length π. The initial temperature distribution is $f(x) = x$, and the ends are held at the fixed temperature 0. Find the heat distribution $u(x, t)$ over time.

(a) $u(x, t) = \sum_{j=1}^{\infty} e^{-a^2 j^2 t} \sin jx$

(b) $u(x, t) = \sum_{j=1}^{\infty} a^2 t^2 \sin jx$

(c) $u(x, t) = \sum_{j=1}^{\infty} e^{a^2 j^2 t} \sin jx$

(d) $u(x, t) = \sum_{j=1}^{\infty} e^{-ajt} \sin jx$

(e) $u(x, t) = \sum_{j=1}^{\infty} e^{-a^2 j^2 t} \dfrac{2}{j} (-1)^{j+1} \sin jx$

60. The Laplace transform of the function $f(x) = xe^x$ is

(a) $L[f](p) = \dfrac{1}{(-p+1)^2}$

(b) $L[f](p) = \dfrac{1}{p^2}$

(c) $L[f](p) = \dfrac{-1}{(p+1)^2}$

(d) $L[f](p) = \dfrac{p}{p^2+1}$

(e) $L[f](p) = \dfrac{-p}{1-p^2}$

61. The Laplace transform of the function $f(x) = 4x^3 - 5\sin 2x$ is

(a) $L[f](p) = \dfrac{8}{p^2} - \dfrac{6}{p^2+1}$

(b) $L[f](p) = \dfrac{1}{p^2-1} + \dfrac{2}{p^2}$

(c) $L[f](p) = \dfrac{p}{p^2+9} - \dfrac{1}{p^2}$

(d) $L[f](p) = \dfrac{24}{p^4} - \dfrac{10}{p^2+4}$

(e) $L[f](p) = \dfrac{16}{p^3} + \dfrac{8}{p^2+1}$

62. The inverse Laplace transform of the function $F(p) = \dfrac{-p^3 + p^2 + p + 4}{p^4 + 5p^2 + 4}$ is

(a) $f(x) = \sin 3x + \sin x$

(b) $f(x) = 4\cos 2x - 3\sin 4x$

(c) $f(x) = \sin x - \cos 2x$

(d) $f(x) = 5\cos 4x + 6\sin 2x$

(e) $f(x) = \cos x + \sin 2x$

63. The inverse Laplace transform of the function $F(p) = \dfrac{-p^4 + 6p - 12}{p^5 - 2p^4}$ is

    (a) $f(x) = x^5 - e^x$

    (b) $f(x) = \sinh x + \cos 2x$

    (c) $f(x) = \sin 3x - e^{4x}$

    (d) $f(x) = x^3 - e^{2x}$

    (e) $f(x) = x^4 + e^{-5x}$

64. Use the Laplace transform to find the general solution of the differential equation $y'' - 4y' + 4y = e^x$.

    (a) $y = Ae^{2x} + Bxe^{2x} + e^x$

    (b) $y = Ae^x + Be^{-x} + 3e^x$

    (c) $y = Ae^{2x} + Be^{-2x} + e^x$

    (d) $y = Ae^x + Bxe^x + e^{2x}$

    (e) $y = Ae^{3x} + Be^x + e^{-x}$

65. Use the Laplace transform to find the general solution of the differential equation $y'' - 4y = x$.

    (a) $y = \dfrac{x}{2} + Ae^x + Be^{-x}$

    (b) $y = \dfrac{-x^2}{2} + Ae^{3x} + Be^{-3x}$

    (c) $y = \dfrac{x}{4} + Ae^{-x} + Be^{-2x}$

    (d) $y = \dfrac{-x}{3} + Ae^{2x} + Be^{3x}$

    (e) $y = \dfrac{-x}{4} + Ae^{2x} + Be^{-2x}$

66. Use the Laplace transform to solve the initial value problem $y' - y = e^x$, $y(0) = 2$.

    (a) $y = e^{-x} + 4$

    (b) $y = xe^x + 2e^x$

    (c) $y = x\sin x - \cos x$

    (d) $y = xe^{-x} - e^x$

    (e) $y = x^2\cos x - x\sin x$

67. Use the Laplace transform to solve the initial value problem $y'' - 5y' + 6y = x$, $y(0) = 1$, $y'(0) = 4$.

    (a) $y = \dfrac{x}{6} + \dfrac{5}{36} + Ae^{2x} + Be^{3x}$

    (b) $y = \dfrac{x^2}{3} - \dfrac{2}{3} + Ae^x + Be^{-2x}$

(c) $y = \dfrac{x-5}{3} + A\cos x + B\sin x$

(d) $y = \dfrac{x^2 - x}{4} + Ax\cos x + Bx\sin x$

(e) $y = \dfrac{x-5}{7} + Ae^{3x} + Be^{-x}$

68. Let $f(x) = \cos x$ and $g(x) = \sin x$. Calculate $L[f * g](p)$.

(a) $L[f * g](p) = \dfrac{1}{p} \cdot \dfrac{p}{4 + p^2}$

(b) $L[f * g](p) = \dfrac{4}{p^2 - 4} \cdot \dfrac{4p}{p^2 - 4}$

(c) $L[f * g](p) = \dfrac{1}{p^2 + 1} \cdot \dfrac{p}{p^2 + 1}$

(d) $L[f * g](p) = \dfrac{2}{p^2} \cdot \dfrac{p}{p + 1}$

(e) $L[f * g](p) = \dfrac{6}{p^2 - 1} \cdot \dfrac{4p}{p^2 + 1}$

69. Use the Laplace transform to solve the integral equation

$$y(x) = x + \int_0^x \sin(x - t)y(t)\, dt.$$

(a) $y = x + \dfrac{x^3}{6}$

(b) $y = -x + \dfrac{x^2}{4}$

(c) $y = x^2 - \dfrac{x}{5}$

(d) $y = x^3 + \dfrac{x^2}{6}$

(e) $y = \dfrac{x}{2} - x^3$

70. Use the Laplace transform to solve the integral equation

$$y(x) = x - \int_0^x e^{x-t}y(t)\, dt.$$

(a) $y = x^2 - \dfrac{x^3}{3}$

(b) $y = x - \dfrac{x^2}{2}$

(c) $y = \dfrac{x}{3} - \dfrac{x^2}{2}$

(d) $y = 2x + 3x^3$

(e) $y = \dfrac{x^3}{5} - \dfrac{x}{4}$

71. Use the Laplace transform to solve the integral equation

$$y(x) = x^2 + \int_0^x (x - t) y(t)\, dt.$$

(a) $y(x) = -2x + e^{2x}$

(b) $y(x) = -2 + e^x + e^{-x}$

(c) $y(x) = x^2 - e^{2x} + e^{-x}$

(d) $y(x) = x + e^x$

(e) $y(x) = -x + e^{-x}$

72. Find the kernel A, coming from the principle of superposition as in Example 6.12, for the differential equation $y'' - y' - 6y = e^x$.

(a) $A(x) = \frac{1}{3} - e^{-3x} + e^x$

(b) $A(x) = \dfrac{x}{4} + e^{2x} - e^{-3x}$

(c) $A(x) = \frac{-2}{3} + e^{-x} - e^{5x}$

(d) $A(x) = \frac{-1}{6} + \frac{1}{15}e^{3x} + \frac{1}{10}e^{-2x}$

(e) $A(x) = \frac{4}{5} - 3e^{2x} - 5e^x$

73. Use the principle of superposition to solve the initial value problem $y'' - 3y' + 2y = e^{3x}$, $y(0) = 2$, $y'(0) = 3$.

(a) $y = e^{3x} - 3e^x + 2e^{2x}$

(b) $y = 4e^{3x} + e^x - 5e^{2x}$

(c) $y = -e^{3x} - e^x + e^{2x}$

(d) $y = \frac{3}{2}e^x + \frac{1}{2}e^{3x}$

(e) $y = e^{-x} + e^{2x} + e^{3x}$

74. Find the Laplace transform of the step function

$$f(x) = \begin{cases} 0 & \text{if } x \le 2 \\ 1 & \text{if } x > 2. \end{cases}$$

(a) $L[f](p) = -\dfrac{1}{p}e^{-2p}$

(b) $L[f](p) = \dfrac{1}{p^2}e^{-p}$

(c) $L[f](p) = \dfrac{1}{p}e^p$

(d) $L[f](p) = \dfrac{2}{p^3}e^{-3p}$

(e) $L[f](p) = \dfrac{1}{p^2}e^p$

75. Let δ be the impulse function and define, for $a > 0$, $p_a(x) = \delta(x - a)$. What is the Laplace transform of p_a?

(a) $L[f](p) = e^{-pa}$

(b) $L[f](p) = e^{pa}$

(c) $L[f](p) = pe^{-pa}$

(d) $L[f](p) = pe^{pa}$

(e) $L[f](p) = p^2 e^{pa}$

76. Calculate the convolution of $f(x) = \cos x$ and $g(x) = x$.

(a) $f * g(x) = \sin x + x$

(b) $f * g(x) = \sin 2x - \cos x$

(c) $f * g(x) = 1 - \cos x$

(d) $f * g(x) = 1 + \cos x$

(e) $f * g(x) = x \cos x - \sin x$

77. Calculate the convolution of $f(x) = x$ with $g(x) = x$.

(a) $f * g(x) = x^2 - x$

(b) $f * g(x) = x^2 + x$

(c) $f * g(x) = x^2 - x^3$

(d) $f * g(x) = \dfrac{x^3}{6}$

(e) $f * g(x) = \dfrac{x^3}{4}$

78. Calculate the convolution of $f(x) = x^2$ and $g(x) = x$.

(a) $f * g(x) = \dfrac{x^3}{9}$

(b) $f * g(x) = \dfrac{x^2}{16}$

(c) $f * g(x) = \dfrac{x}{3}$

(d) $f * g(x) = \dfrac{x + x^2}{12}$

(e) $f * g(x) = \dfrac{x^4}{12}$

79. Carry out three iterations of Euler's method for the differential equation $y' = x + 2y$ with initial condition $y(0) = 2$ and step size 0.2.
    (a) 4.78
    (b) 5.624
    (c) 5.33
    (d) 4.94
    (e) 5.1

80. Carry out four iterations of Euler's method for the differential equation $y' = 2x - y$ with initial condition $y(0) = -1$ and step size 0.1.
    (a) −0.5465
    (b) 0.4976
    (c) −0.6983
    (d) −0.5439
    (e) −0.5598

81. Carry out three iterations of the improved Euler's method for the differential equation $y' = x + 2y$ with initial condition $y(0) = 2$ and step size 0.2.
    (a) 6.24179
    (b) 5.997348
    (c) 6.742552
    (d) 5.902857
    (e) 6.114976

82. Carry out four iterations of the improved Euler's method for the differential equation $y' = 2x - y$ with initial condition $y(0) = -1$ and step size 0.1.
    (a) −0.556792
    (b) −0.638044
    (c) −0.563056
    (d) 0.529119
    (e) −0.529198

83. The most convenient method for telling when the result of a numerical method is accurate to m decimal places is
    (a) Iterate the method for m steps.
    (b) Use a method that is known to double its accuracy with each iteration, and then apply it $[\log_2 m]$ times.
    (c) Iterate the method until the solution has m significant figures.
    (d) Use scientific notation.
    (e) Iterate the method until the kth step and the $(k+1)$st step agree to m decimal places. Then use the result from the $(k+1)$st step.

84. Carry out two iterations of the Runge–Kutta method for the differential equation $y' = x + 2y$ with initial condition $y(0) = 2$ and step size 0.2.
    (a) 4.566827
    (b) 4.678398
    (c) 5.129845
    (d) 4.998127
    (e) −0.546789

85. Carry out two iterations of the Runge–Kutta method for the differential equation $y' = 2x - y$ with initial condition $y(0) = -1$ and step size 0.1.
    (a) 0.84673
    (b) −0.78349
    (c) −0.85792
    (d) −0.88923
    (e) 0.77889

86. The system

$$\begin{cases} \dfrac{dx}{dt} = x + 2y \\ \dfrac{dy}{dt} = -x + 4y \end{cases}$$

has the two solution sets

(a) $\begin{cases} x = e^{2t} \\ y = e^{2t} \end{cases}$ and $\begin{cases} x = 3e^{3t} \\ y = e^{3t} \end{cases}$

(b) $\begin{cases} x = e^{3t} \\ y = e^{3t} \end{cases}$ and $\begin{cases} x = 2e^{2t} \\ y = e^{2t} \end{cases}$

(c) $\begin{cases} x = e^{-2t} \\ y = e^{-2t} \end{cases}$ and $\begin{cases} x = 3e^{-3t} \\ y = e^{-3t} \end{cases}$

(d) $\begin{cases} x = e^{t} \\ y = e^{t} \end{cases}$ and $\begin{cases} x = 4e^{-t} \\ y = e^{-t} \end{cases}$

(e) $\begin{cases} x = 3e^{t} \\ y = e^{3t} \end{cases}$ and $\begin{cases} x = 2e^{t} \\ y = e^{-2t} \end{cases}$

87. The system

$$\begin{cases} \dfrac{dx}{dt} = -2x + 4y \\ \dfrac{dy}{dt} = x + y \end{cases}$$

has the two solution sets

(a) $\begin{cases} x = e^t \\ y = e^{2t} \end{cases}$ and $\begin{cases} x = 3e^{-3t} \\ y = e^{-4t} \end{cases}$

(b) $\begin{cases} x = e^{2t} \\ y = 4e^{2t} \end{cases}$ and $\begin{cases} x = 2e^{-2t} \\ y = e^{-2t} \end{cases}$

(c) $\begin{cases} x = e^{2t} \\ y = e^{-2t} \end{cases}$ and $\begin{cases} x = 3e^{-3t} \\ y = e^{3t} \end{cases}$

(d) $\begin{cases} x = e^t \\ y = e^{-t} \end{cases}$ and $\begin{cases} x = 3e^t \\ y = e^{-t} \end{cases}$

(e) $\begin{cases} x = -4e^{-3t} \\ y = e^{-3t} \end{cases}$ and $\begin{cases} x = e^{2t} \\ y = e^{2t} \end{cases}$

88. The system

$$\begin{cases} \dfrac{dx}{dt} = 3x - 5y \\[2mm] \dfrac{dy}{dt} = 2x + y \end{cases}$$

has the two solution sets

(a) $\begin{cases} x = e^{3t} \cos 2t + 2e^{3t} \sin 2t \\ y = 2e^{3t} \sin 2t \end{cases}$ and $\begin{cases} x = 3e^{-3t} \cos 2t - 4e^{-3t} \sin 2t \\ y = e^{-3t} \cos 2t + 2e^{-3t} \sin 2t \end{cases}$

(b) $\begin{cases} x = e^{-t} \cos 3t + 2e^t \sin 3t \\ y = 2e^{-3t} \sin 2t \end{cases}$ and $\begin{cases} x = 2e^t \cos t - 4e^{-t} \sin t \\ y = e^t \cos t + 2e^{-t} \sin t \end{cases}$

(c) $\begin{cases} x = e^{2t} \cos 2t + 2e^{2t} \sin 2t \\ y = 2e^{2t} \sin 2t - 3e^{2t} \cos 2t \end{cases}$ and $\begin{cases} x = 3e^t \cos 4t - 4e^t \sin 4t \\ y = e^{-3t} \cos 4t + 2e^{-3t} \sin 4t \end{cases}$

(d) $\begin{cases} x = e^{-2t} \cos t + 2e^{-2t} \sin t \\ y = 2e^{-t} \sin t + 4e^{-t} \cos t \end{cases}$ and $\begin{cases} x = 3e^{-3t} \cos 2t - 4e^{-3t} \sin 2t \\ y = e^{-3t} \cos 2t + 2e^{-3t} \sin 2t \end{cases}$

(e) $\begin{cases} x = e^{2t} \cos 3t - 3e^{2t} \sin 3t \\ y = 2e^{2t} \cos 3t \end{cases}$ and $\begin{cases} x = 3e^{2t} \cos 3t + e^{2t} \sin 3t \\ y = 2e^{2t} \sin 3t \end{cases}$

89. The system

$$\begin{cases} \dfrac{dx}{dt} = 2x \\[2mm] \dfrac{dy}{dt} = 3y \end{cases}$$

is not interesting (from the point of view of the methods presented in Chapter 8) because

(a) the only solution set is $x \equiv 0$, $y \equiv 0$.

(b) the system is uncoupled and the equations may be solved individually.

(c) all the solutions are linearly dependent.

(d) the two differential equations are inconsistent.

(e) the equations are nonlinear.

90. The system

$$\begin{cases} \dfrac{dx}{dt} = 2x - 4y \\[2mm] \dfrac{dy}{dt} = 3x + 2y \end{cases}$$

has the two solution sets

(a) $\begin{cases} x = 2e^{-2t}\cos 2t \\ y = 3e^{2t}\sin 2t \end{cases}$ and $\begin{cases} x = 3e^{-2t}\cos 2t \\ y = e^{2t}\sin 2t \end{cases}$

(b) $\begin{cases} x = e^{-t}\cos 3\sqrt{2}t \\ y = 2e^{-t}\sin 3\sqrt{2}t \end{cases}$ and $\begin{cases} x = 2e^{t}\cos 3\sqrt{2}t \\ y = e^{t}\sin 3\sqrt{2}t \end{cases}$

(c) $\begin{cases} x = 4e^{2t}\cos 3\sqrt{2}t \\ y = 2e^{2t}\sin 3\sqrt{2}t \end{cases}$ and $\begin{cases} x = -5e^{2t}\cos 3\sqrt{2}t \\ y = 2e^{2t}\sin 3\sqrt{2}t \end{cases}$

(d) $\begin{cases} x = 2e^{2t}\cos 2\sqrt{3}t \\ y = \sqrt{3}e^{2t}\sin 2\sqrt{3}t \end{cases}$ and $\begin{cases} x = 2e^{2t}\sin 2\sqrt{3}t \\ y = -\sqrt{3}e^{2t}\cos 2\sqrt{3}t \end{cases}$

(e) $\begin{cases} x = e^{-4t}\cos \sqrt{3}t \\ y = 2e^{-4t}\sin \sqrt{3}t \end{cases}$ and $\begin{cases} x = 3e^{-3t}\cos 2t - 4e^{-3t}\sin 2t \\ y = e^{-3t}\cos 2t + 2e^{-3t}\sin 2t \end{cases}$

91. The system

$$\begin{cases} \dfrac{dx}{dt} = -x + y \\[2mm] \dfrac{dy}{dt} = -4x + 3y \end{cases}$$

has the two solution sets

(a) $\begin{cases} x = e^{t} \\ y = 2e^{t} \end{cases}$ and $\begin{cases} x = te^{t} \\ y = (1 + 2t)e^{t} \end{cases}$

(b) $\begin{cases} x = 2e^{-t}\cos 2t \\ y = 3e^{t}\sin 2t \end{cases}$ and $\begin{cases} x = e^{2t} \\ y = e^{-2t} \end{cases}$

(c) $\begin{cases} x = (t + 1)e^{2t} \\ y = (t - 1)e^{-2t} \end{cases}$ and $\begin{cases} x = e^{-3t} \\ y = e^{-t} \end{cases}$

(d) $\begin{cases} x = 2te^{4t} \\ y = 3te^{4t} \end{cases}$ and $\begin{cases} x = 4e^{3t} \\ y = -4e^{-3t} \end{cases}$

(e) $\begin{cases} x = e^{t} \\ y = 2e^{-t} \end{cases}$ and $\begin{cases} x = te^{t} \\ y = te^{-t} \end{cases}$

92. The system

$$\begin{cases} \dfrac{dx}{dt} = 4x + y \\[2mm] \dfrac{dy}{dt} = -x + 2y \end{cases}$$

has the two solution sets

(a) $\begin{cases} x = e^t \\ y = 2e^t \end{cases}$ and $\begin{cases} x = te^{2t} \\ y = (1 + 2t)e^{2t} \end{cases}$

(b) $\begin{cases} x = 2e^{-t} \\ y = 3e^{-t} \end{cases}$ and $\begin{cases} x = e^{2t} \\ y = e^{-2t} \end{cases}$

(c) $\begin{cases} x = (t + 1)e^{2t} \\ y = (t - 1)e^{-2t} \end{cases}$ and $\begin{cases} x = e^{-3t} \\ y = e^{-t} \end{cases}$

(d) $\begin{cases} x = e^{3t} \\ y = -e^{3t} \end{cases}$ and $\begin{cases} x = (1 + t)e^{3t} \\ y = -te^{3t} \end{cases}$

(e) $\begin{cases} x = te^{3t} \\ y = 2te^{3t} \end{cases}$ and $\begin{cases} x = 4e^{3t} \\ y = -e^{3t} \end{cases}$

93. The differential equation $y'' - 2xy' + 3xy = 0$ is equivalent to the system

(a) $\begin{cases} \dfrac{dy}{dx} = -xz \\[2mm] \dfrac{dz}{dx} = 2z - 4xy \end{cases}$

(b) $\begin{cases} \dfrac{dy}{dx} = 4xz + y \\[2mm] \dfrac{dz}{dx} = -xz + 2xy \end{cases}$

(c) $\begin{cases} \dfrac{dy}{dx} = z \\[2mm] \dfrac{dz}{dx} = 2xz - 3xy \end{cases}$

(d) $\begin{cases} \dfrac{dy}{dx} = xy - z \\[2mm] \dfrac{dz}{dx} = xz - 3y \end{cases}$

(e) $\begin{cases} \dfrac{dy}{dx} = 4z - 2xy \\[2mm] \dfrac{dz}{dx} = xz + 3y \end{cases}$

94. The differential equation $y''' + 4xy'' - xy' + 3y = x^2$ is equivalent to the system

(a)
$$\begin{cases} \dfrac{dy}{dx} = z \\[2mm] \dfrac{dz}{dx} = w \\[2mm] \dfrac{dw}{dx} = -4xw + xz - 3y + x^2 \end{cases}$$

(b)
$$\begin{cases} \dfrac{dy}{dx} = z - w \\[2mm] \dfrac{dz}{dx} = w + y \\[2mm] \dfrac{dw}{dx} = -xw + z - 2y + x^2 \end{cases}$$

(c)
$$\begin{cases} \dfrac{dy}{dx} = w - y \\[2mm] \dfrac{dz}{dx} = z - w \\[2mm] \dfrac{dw}{dx} = -xz + xy - 3w + x \end{cases}$$

(d)
$$\begin{cases} \dfrac{dy}{dx} = w \\[2mm] \dfrac{dz}{dx} = z \\[2mm] \dfrac{dw}{dx} = -4xy + xw - 3z \end{cases}$$

(e)
$$\begin{cases} \dfrac{dy}{dx} = y \\[2mm] \dfrac{dz}{dx} = z \\[2mm] \dfrac{dw}{dx} = -4xw - 3y \end{cases}$$

95. The differential equation $(y'')^2 - (y')^3 - y^4 = x^2$ is equivalent to the system

(a)
$$\begin{cases} (z')^2 - z^3 - y^4 = x^2 \\ y' = z \end{cases}$$

(b)
$$\begin{cases} (z') - z^2 - y^3 = x^2 \\ y' = z \end{cases}$$

(c)
$$\begin{cases} (z')^2 - z^2 - y^3 = x^2 \\ y' = z^2 \end{cases}$$

(d) $\begin{cases} z'y - z^3 y' - y^4 = x^2 \\ y'z = x \end{cases}$

(e) $\begin{cases} (z')^4 - z^3 - zy^4 = x^2 \\ y'x = z \end{cases}$

96. The system

$$\begin{cases} \dfrac{dx}{dt} = x + 2y + 2t \\[2mm] \dfrac{dy}{dt} = -x + 4y + 3t^2 \end{cases}$$

has the general solution

(a) $\begin{cases} x = Ae^{2t} + 2Be^{3t} + t^2 + \dfrac{t}{5} + \dfrac{7}{18} \\[2mm] y = Ae^{2t} + Be^{3t} - \dfrac{t^2}{3} - \dfrac{t}{4} + \dfrac{5}{36} \end{cases}$

(b) $\begin{cases} x = Ae^{t} + 2Be^{-2t} + 2t^2 + \dfrac{t}{5} + \dfrac{5}{16} \\[2mm] y = Ae^{t} + Be^{-2t} - \dfrac{t^2}{4} - \dfrac{t}{2} + \dfrac{1}{36} \end{cases}$

(c) $\begin{cases} x = Ae^{-t} + 2Be^{-t} + 4t^2 + \dfrac{2t}{3} + \dfrac{1}{18} \\[2mm] y = Ae^{-3t} + Be^{t} - \dfrac{t^2}{4} - \dfrac{t}{9} + \dfrac{11}{36} \end{cases}$

(d) $\begin{cases} x = 3Ae^{3t} + 2Be^{2t} + 2t^2 + \dfrac{t}{3} + \dfrac{1}{18} \\[2mm] y = 5Ae^{3t} + Be^{2t} - \dfrac{t^2}{3} - \dfrac{t}{6} + \dfrac{5}{36} \end{cases}$

(e) $\begin{cases} x = Ae^{3t} + 2Be^{2t} + t^2 + \dfrac{t}{3} + \dfrac{5}{18} \\[2mm] y = Ae^{3t} + Be^{2t} - \dfrac{t^2}{2} - \dfrac{t}{6} + \dfrac{1}{36} \end{cases}$

97. The system

$$\begin{cases} \dfrac{dx}{dt} = -2x + 4y + \cos t \\[2mm] \dfrac{dy}{dt} = x + y + \sin t \end{cases}$$

has the general solution

(a)
$$\begin{cases} x = -4Ae^{-3t} + Be^{2t} + \frac{2}{25}\cos t - \frac{11}{25}\sin t \\ y = Ae^{-3t} + Be^{2t} - \frac{8}{25}\cos t - \frac{6}{25}\sin t \end{cases}$$

(b)
$$\begin{cases} x = -2Ae^{-3t} + 3Be^{2t} + \frac{1}{25}\cos t - \frac{9}{25}\sin t \\ y = 3Ae^{-3t} + 2Be^{2t} - \frac{4}{25}\cos t - \frac{3}{25}\sin t \end{cases}$$

(c)
$$\begin{cases} x = -Ae^{-3t} + 2Be^{2t} + \frac{3}{25}\cos t - \frac{7}{25}\sin t \\ y = 3Ae^{-3t} + 5Be^{2t} - \frac{2}{25}\cos t - \frac{3}{25}\sin t \end{cases}$$

(d)
$$\begin{cases} x = 2Ae^{-t} + -3Be^{2t} + \frac{6}{25}\cos t - \frac{2}{25}\sin t \\ y = 2Ae^{-t} + 4Be^{2t} - \frac{4}{25}\cos t - \frac{3}{25}\sin t \end{cases}$$

(e)
$$\begin{cases} x = -4Ae^{t} + Be^{-2t} + \frac{7}{25}\cos t - \frac{2}{25}\sin t \\ y = Ae^{t} + Be^{-2t} - \frac{3}{25}\cos t - \frac{4}{25}\sin t \end{cases}$$

98. The system

$$\begin{cases} \dfrac{dx}{dt} = 3x - 5y + e^{t} \\ \dfrac{dy}{dt} = 2x + y + 2e^{t} \end{cases}$$

has the general solution

(a)
$$\begin{cases} x = A[e^{4t}\cos 5t - 3e^{-4t}\sin 5t] + B[3e^{4t}\cos 5t + e^{-4t}\sin 5t] - 6e^{t} \\ y = A[2e^{4t}\cos 5t] + B[2e^{-4t}\sin 5t] - \frac{1}{3}e^{t} \end{cases}$$

(b)
$$\begin{cases} x = A[e^{-2t}\cos 3t - 3e^{-3t}\sin 2t] + B[3e^{2t}\cos 3t + e^{2t}\sin 3t] - e^{2t} \\ y = A[2e^{2t}\cos 3t] + B[2e^{3t}\sin 2t] - \frac{1}{5}e^{2t} \end{cases}$$

(c)
$$\begin{cases} x = A[e^{-t}\cos t - 3e^{-t}\sin t] + B[3e^{-t}\cos t + e^{-t}\sin t] - 4e^{t} \\ y = A[2e^{t}\cos t] + B[2e^{t}\sin t] - \frac{1}{2}e^{t} \end{cases}$$

(d)
$$\begin{cases} x = A[e^{2t}\cos 3t - 3e^{2t}\sin 3t] + B[3e^{2t}\cos 3t + e^{2t}\sin 3t] - e^{t} \\ y = A[2e^{2t}\cos 3t] + B[2e^{2t}\sin 3t] - \frac{1}{5}e^{t} \end{cases}$$

(e)
$$\begin{cases} x = A[e^{t}\cos 2t - 3e^{t}\sin 2t] + B[3e^{2t}\cos 3t + e^{2t}\sin 3t] - 2e^{t} \\ y = A[2e^{t}\cos 2t] + B[2e^{t}\sin 2t] - \frac{1}{8}e^{t} \end{cases}$$

99. The system

$$\begin{cases} \dfrac{dx}{dt} = 2x - 4y + 3 \\ \dfrac{dy}{dt} = 3x + 2y - t^2 \end{cases}$$

has the general solution

(a)
$$\begin{cases} x = A[2e^t \cos \sqrt{3}t] + B[2e^t \sin \sqrt{3}t] + \dfrac{t^2}{3} - \dfrac{t}{6} - \dfrac{1}{8} \\ y = A[\sqrt{5}e^{3t} \sin \sqrt{3}t] + B[-\sqrt{5}e^{3t} \cos \sqrt{3}t] + \dfrac{t^2}{4} - \dfrac{t}{2} + \dfrac{13}{32} \end{cases}$$

(b)
$$\begin{cases} x = A[e^{2t} \cos 2\sqrt{7}t] + B[e^{2t} \sin 2\sqrt{7}t] + \dfrac{t^2}{4} - \dfrac{3}{8} \\ y = A[\sqrt{6}e^{2t} \sin \sqrt{7}t] + B[-\sqrt{5}e^{2t} \cos \sqrt{7}t] + \dfrac{t^2}{3} - \dfrac{t}{12} \end{cases}$$

(c)
$$\begin{cases} x = A[2e^{2t} \cos 2\sqrt{3}t] + B[2e^{2t} \sin 2\sqrt{3}t] + \dfrac{t^2}{4} + \dfrac{t}{8} - \dfrac{3}{8} \\ y = A[\sqrt{3}e^{2t} \sin 2\sqrt{3}t] + B[-\sqrt{3}e^{2t} \cos 2\sqrt{3}t] + \dfrac{t^2}{8} - \dfrac{t}{16} + \dfrac{17}{32} \end{cases}$$

(d)
$$\begin{cases} x = A[e^{-2t} \cos 3t - 3e^{-2t} \sin 3t] + B[3e^{2t} \cos 2t + e^{2t} \sin 2t] - 3e^t \\ y = A[2e^{-2t} \cos 3t] + B[2e^{2t} \sin 3t] - \tfrac{1}{4}e^t \end{cases}$$

(e)
$$\begin{cases} x = A[e^{-3t} \cos t - 3e^{-3t} \sin t] + B[3e^{3t} \cos 3t + e^{3t} \sin 3t] - e^t \\ y = A[2e^t \cos 2t] + B[2e^t \sin 2t] - \tfrac{1}{6}e^t \end{cases}$$

100. The system

$$\begin{cases} \dfrac{dx}{dt} = -x + y - \sin t \\ \dfrac{dy}{dt} = -4x + 3y + \cos t \end{cases}$$

has the general solution

(a)
$$\begin{cases} x = A[e^{2t}] + B[(1 + 4t)e^{3t}] + \tfrac{3}{4} \cos t \\ y = A[5e^{2t}] + B[-2te^{3t}] + \tfrac{3}{4} \cos t - \tfrac{1}{3} \sin t \end{cases}$$

(b)
$$\begin{cases} x = A[e^{3t}] + B[(1 + t)e^{3t}] + \tfrac{3}{2} \cos t \\ y = A[-e^{3t}] + B[-te^{3t}] + \tfrac{3}{2} \cos t - \tfrac{1}{2} \sin t \end{cases}$$

(c)
$$\begin{cases} x = A[e^t] + B[(3-t)e^{-t}] + \sin t \\ y = A[-e^t] + B[-te^t] + \frac{3}{4}\cos t - \frac{2}{3}\cos t \end{cases}$$

(d)
$$\begin{cases} x = A[e^{2t}] + B[te^{-2t}] - \cos t \\ y = A[-4e^{3t}] + B[te^{-3t}] + \frac{3}{5}\cos t - \sin t \end{cases}$$

(e)
$$\begin{cases} x = A[e^t] + B[(-1+3t)e^t] + \frac{1}{4}\cos t \\ y = A[-(1+t)e^t] + B[te^{-t}] - \frac{1}{3}\sin t \end{cases}$$

Solutions

1. (a), 2. (c), 3. (d), 4. (a), 5. (b), 6. (e), 7. (b), 8. (d), 9. (e), 10. (c),
11. (c), 12. (c), 13. (a), 14. (e), 15. (c), 16. (d), 17. (e), 18. (d), 19. (b), 20. (a),
21. (a), 22. (d), 23. (c), 24. (b), 25. (b), 26. (d), 27. (e), 28. (b), 29. (a), 30. (a),
31. (c), 32. (b), 33. (c), 34. (e), 35. (a), 36. (e), 37. (d), 38. (b), 39. (c), 40. (a),
41. (b), 42. (d), 43. (e), 44. (d), 45. (c), 46. (e), 47. (c), 48. (b), 49. (a), 50. (d),
51. (a), 52. (e), 53. (b), 54. (c), 55. (b), 56. (a), 57. (c), 58. (d), 59. (e), 60. (a),
61. (d), 62. (c), 63. (d), 64. (a), 65. (e), 66. (b), 67. (a), 68. (c), 69. (a), 70. (b),
71. (b), 72. (d), 73. (d), 74. (a), 75. (a), 76. (c), 77. (d), 78. (e), 79. (b), 80. (d),
81. (c), 82. (e), 83. (e), 84. (a), 85. (c), 86. (b), 87. (e), 88. (e), 89. (b), 90. (d),
91. (a), 92. (d), 93. (c), 94. (a), 95. (a), 96. (e), 97. (a), 98. (d), 99. (c), 100. (b).

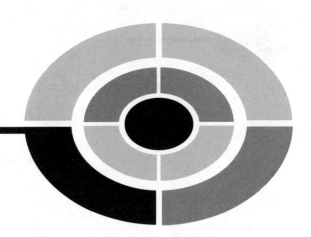

Solutions to Exercises

Chapter 1

1. (a) $y' = (x^2 + c)' = 2x$

   (b) $xy' = x \cdot (cx^2)' = x \cdot 2cx = 2cx^2 = 2y$

2. (a) $y = \int e^{3x} - x\,dx = \dfrac{e^{3x}}{3} - \dfrac{x^2}{2} + C$

   (b) $y = \int xe^{x^2}\,dx = \dfrac{e^{x^2}}{2} + C$

3. (a) $y = \int xe^x\,dx = xe^x - e^x + C$. The initial condition says that $3 = y(1) = e - e + C$, hence $C = 3$. The solution to the initial value problem is $y = xe^x - e^x + 3$.

   (b) $y = \int 2\sin x \cos x\,dx = \sin^2 x + C$. The initial condition says that $1 = y(0) = 0 + C$, hence $C = 1$. The solution to the initial value problem is $y = \sin^2 x + 1$.

4. (a) Write the equation as

$$x^5 \frac{dy}{dx} - \frac{1}{y^5} = 0.$$

This can be rewritten as

$$y^5 \, dy = x^{-5} \, dx$$

or

$$\int y^5 \, dy = \int x^{-5} \, dx.$$

Integrating out yields

$$\frac{y^6}{6} = -\frac{x^{-4}}{4} + C$$

or

$$y = \left[-\frac{3}{2} x^{-4} + C \right]^{1/6}.$$

(b) Write the equation as

$$\frac{dy}{dx} = 4xy,$$

hence

$$\frac{dy}{y} = 4x \, dx.$$

Integrating both sides gives

$$\ln y = 2x^2 + C$$

or

$$y = De^{x^2}.$$

5. (a) Write the equation as

$$y \, dy = (x + 1) \, dx.$$

Integrating both sides yields

$$\frac{y^2}{2} = \frac{x^2}{2} + x + C.$$

A little algebra then gives

$$y = \sqrt{x^2 + 2x + D}.$$

The initial condition yields

$$3 = y(1) = \sqrt{3 + D},$$

hence $D = 6$. The solution to the initial value problem is

$$y = \sqrt{x^2 + 2x + 6}.$$

(b) We write

$$\frac{dy}{y} = \frac{dx}{x^2}.$$

Integrating yields

$$\ln y = -\frac{1}{x} + C.$$

Exponentiation gives

$$y = De^{-1/x}.$$

The initial condition tells us that

$$2 = De^{-1},$$

hence

$$D = 2e.$$

The solution of the initial value problem is

$$y = 2e \cdot e^{-1/x}.$$

6. (a) Now $\int p(x)\,dx = -x^2/2$, so our integrating factor is $e^{-x^2/2}$. Thus the equation becomes

$$e^{-x^2/2}y' - e^{-x^2/2}xy = 0$$

or

$$\left[e^{-x^2/2}y\right]' = 0$$

or (integrating)

$$e^{-x^2/2}y = C.$$

We find, then, that the general solution is

$$y = Ce^{x^2/2}.$$

(b) Now $\int p(x)\,dx = x^2$, hence our integrating factor is e^{x^2}. Thus the equation becomes

$$e^{x^2}y' + e^{x^2}2xy = e^{x^2}2x$$

or

$$\left[e^{x^2}y\right]' = e^{x^2}2x$$

or (integrating)

$$e^{x^2}y = e^{x^2} + C.$$

Simplifying yields

$$y = 1 + Ce^{-x^2}.$$

7. Let $B(t)$ be the amount of salt in the tank at time t. The initial condition is $B(0) = 2$. The rate of change of the amount of salt present is dB/dt. Since 3 pounds of salt are added per minute, we have a factor of $+3$ on the right-hand side. Since 4 gallons of the mixture are removed, we have a factor of $4B/(10-t)$ on the right-hand side. Thus our differential equation is

$$\frac{dB}{dt} = 3 - 4 \cdot \frac{B}{10-t}.$$

This is easily solved by the method of first-order linear equations to yield

$$B = (10-t)^{-7} + C(10-t)^{-4}.$$

The condition $B(0) = 2$ says that $C = 2 \cdot 10^4 - 10^{-3}$. Thus the amount of salt at any time t is given by

$$B(t) = (10-t)^{-7} + [2 \cdot 10^4 - 10^{-3}] \cdot (10-t)^{-4}.$$

8. (a) Since $[d/dy](x + 2/y) \neq [d/dx](y)$, we see that this equation is not exact.

(b) Since $[d/dy](\sin x \tan y + 1) = \sin x \sec^2 y$ and $[d/dx] \times (-\cos x \sec^2 y) = \sin x \sec^2 y$, we affirm that the equation is exact. Now

$$\int \sin x \tan y + 1 \, dx = -\cos x \tan y + x + \phi(y) = f(x, y).$$

Thus

$$-\cos x \sec^2 y = \frac{\partial f}{\partial y} = -\cos x \sec^2 y + \phi'(y),$$

hence

$$\phi'(y) = 0$$

or $\phi(y) = C$. In sum, $f(x, y) = -\cos x \tan y + x + C$, so the solution of the differential equation is

$$-\cos x \tan y + x = \widetilde{C}.$$

9. Now $dy/dx = 4cx^3$. The negative reciprocal is $-1/[4cx^3]$. Thus the orthogonal trajectories satisfy $dy/dx = -1/[4cx^3]$. Integration yields the family $y = x^{-2}/[8C]$ of orthogonal trajectories.

10. (a) We rewrite the equation as

$$\left(\frac{y}{x} \sin \frac{y}{x} + 1\right) = \sin \frac{y}{x} \frac{dy}{dx}.$$

Replacing y by ty and x by tx reveals the equation to be homogeneous of degree 0. Now we make the substitutions

$$y = zx \quad \text{and} \quad \frac{dy}{dx} = z + x\frac{dz}{dx}$$

to obtain

$$z \sin z + 1 = \sin z \left[z + x\frac{dz}{dx}\right].$$

This equation is easily solved by separation of variables to yield

$$z = \arccos[-\ln x + C].$$

Resubstituting $z = y/x$ finally gives

$$y = x \arccos[-\ln x + C].$$

(b) Writing the equation as

$$\frac{dy}{dx} = \frac{dy}{dx} + 2e^{-y/x},$$

we see that it is homogeneous of degree 0. Substituting $z = y/x$ and $dy/dx = z + x[dz/dx]$, we obtain the equation

$$x\frac{dz}{dx} = 2e^{-z}.$$

This is easily solved using separation of variables to obtain

$$z = \ln[2\ln x + C],$$

hence

$$y = x \cdot \ln[2\ln x + C].$$

11. (a) We calculate that

$$g(x) = \frac{[\partial M/\partial y] - [\partial N/\partial x]}{N} = -\frac{2}{x}.$$

It follows that the integrating factor we seek is $\mu(x) = e^{\int g(x)\,dx} = 1/x^2$. Multiplying the differential equation through by μ gives

$$12y\,dx + 12x\,dy = 0.$$

This equation is certainly exact, and can be solved by the standard method. The answer is

$$12xy = C.$$

(b) We calculate that

$$g(x) = \frac{[\partial M/\partial y] - [\partial N/\partial x]}{N} = -\frac{1}{x}.$$

It follows that the integrating factor we seek is $\mu(x) = e^{\int g(x)\,dx} = e^{-\ln x} = 1/x$. Multiplying the differential equation through by μ gives

$$\left(y - \frac{1}{x}\right)dx + (x - y)\,dy = 0.$$

This equation is exact and may be solved by the usual means. The solution is

$$yx - \ln x - \frac{y^2}{2} = C.$$

12. (a) The change of variable $y' = p$, $y'' = p'$, converts the equation to

$$xp' = p + p^3.$$

Now separation of variables may be used to find that

$$p = \frac{\sqrt{C}x}{\sqrt{1 - Cx^2}}.$$

Resubstituting for y yields the final answer

$$y = -\frac{\sqrt{1 - Cx^2}}{\sqrt{C}}.$$

(b) We use the substitution $y' = p$, $y'' = p\dfrac{dp}{dy}$ to obtain the new equation

$$p\frac{dp}{dy} = k^2 y.$$

This equation is easily solved by separation of variables to obtain a solution

$$y = C \cdot e^{kx}.$$

Chapter 2

1. (a) The associated polynomial is $r^2 + r - 6$ with roots $r = 3, -2$. The general solution to the differential equation is $y = Ae^{3x} + Be^{-2x}$.

(b) The associated polynomial is $r^2 + 2r + 1$. This polynomial has the root 1 repeated. The general solution to the differential equation is $y = e^x + xe^x$.

2. (a) The associated polynomial is $r^2 - 5r + 6$ with roots $r = 2, 3$. The general solution of the differential equation is $y = Ae^{2x} + Be^{3x}$. The initial conditions give $e^2 = Ae^2 + Be^3$ and $3e^2 = 2Ae^2 + 3Be^3$. Solving yields $A = 0$, $B = e^{-1}$. The solution to the initial value problem is $y = e^{-1}e^{3x}$.

(b) The associated polynomial is $r^2 - 6r + 5$ with roots $r = 1, 5$. The general solution of the differential equation is $y = Ae^x + Be^{5x}$. The initial conditions give $3 = A + B$ and $11 = A + 5B$. Solving yields $A = 1$, $B = 2$. The solution to the initial value problem is $y = e^x + 2e^{5x}$.

3. (a) The associated polynomial will be $(r - 1)(r + 2) = r^2 + r - 2$. The differential equation is then $y'' + y' - 2y = 0$.

   (b) The associated polynomial will be $r(r - 2) = r^2 - 2r$. The differential equation is then $y'' - 2y' = 0$.

4. (a) The associated polynomial is $r^2 + 3r - 10$ with roots $r = -5, 2$. The solutions of the homogeneous equation are then $y_1 = e^{-5x}$ and $y_2 = e^{2x}$. We solve the equations

   $$v_1' e^{-5x} + v_2' e^{2x} = 0$$

   $$v_1'(-5e^{-5x}) + v_2'(2e^{2x}) = 6e^{4x}.$$

   The solution is $v_1' = [-6/7]e^{9x}$, $v_2' = [6/7]e^{2x}$. We find then that $y_p = v_1 y_1 + v_2 y_2 = [1/3]e^{4x}$. The general solution of the differential equation is

   $$y = Ae^{-5x} + Be^{2x} + \tfrac{1}{3}e^{4x}.$$

   (b) The associated polynomial is $r^2 + 4$ with roots $r = \pm 2i$. The solutions of the homogeneous equation are then $y_1 = \cos 2x$ and $y_2 = \sin 2x$. We solve the equations

   $$v_1' \cos 2x + v_2' \sin 2x = 0$$

   $$v_1'(-2 \sin 2x) + v_2'(2 \cos 2x) = 3 \sin x.$$

   The solution is $v_1' = [-3/2] \sin x \sin 2x$, $v_2' = [3/2] \sin x \cos 2x$. We find then that $y_p = v_1 y_1 + v_2 y_2 = \sin x$. The general solution of the differential equation is

   $$y = A \cos 2x + B \sin 2x + \sin x.$$

5. (a) Guess a particular solution of the form $y_p = \alpha x + \beta x^2$. Substitute this into the differential equation and solve for the coefficients. The result is $\alpha - 1/3$, $\beta = 1/3$. A particular solution is then $y_p = [-1/3]x + [1/3]x^2$.

   (b) Guess a particular solution of the form $y_p = \alpha x e^{-x} + \beta e^{-x}$. Substitute this into the differential equation and solve for the coefficients. The result is $\alpha = 1/4$, $\beta = 0$. A particular solution is then $y_p = [1/4]x e^x$.

6. (a) Write the equation as

   $$y'' - \frac{2x}{x^2 - 1}y' + \frac{2}{x^2 - 1}y = x^2 + 1. \qquad (*)$$

We guess that $y_1 = x$ is a solution of the homogeneous equation, and this is verified by a quick calculation. Using the method of Section 2.4, we seek another solution of the form $y_2 = v \cdot y_1$. Here

$$v(x) = \int \frac{1}{(y_1)^2} e^{-\int p(x)\, dx}\, dx,$$

where p is the coefficient of y' in (*). We find that $v = x + 1/x$. Thus $y_2 = x^2 + 1$.

Now we guess a function of the form $y_p = \alpha x^2 + \beta x + \gamma$ for a particular solution and find that $y_p = x^2/2 + 1$. Thus the general solution to the differential equation is

$$y = Ax + B\left[x^2 + 1\right] + \frac{x^2}{2} + 1.$$

(b) Write the equation as

$$y'' + \frac{2x+1}{x^2+x}y' + \frac{2x+1}{x(x^2+x)}y = \frac{-4x^2 - 2x}{x^2 + x}. \qquad (\star)$$

We guess that $y_1 = x$ is a solution of the homogeneous equation, and this is verified by a quick calculation. Using the method of Section 2.4, we seek another solution of the form $y_2 = v \cdot y_1$. Here

$$v(x) = \int \frac{1}{(y_1)^2} e^{-\int p(x)\, dx}\, dx,$$

where p is the coefficient of y' in ($\star$). We find that $v = \ln x + 1/x - 1/[2x^2] - \ln(x+1)$. Thus $y_2 = x \ln x + 1 - 1/[2x] - x \ln(x+1)$.

Now we guess a function of the form $y_p = \alpha x^2 + \beta x$ for a particular solution and find that $y_p = -x^2$. Thus the general solution to the differential equation is

$$y = Ax + B\left[x \ln x + 1 - \frac{1}{2x} - x \ln(x+1)\right] - x^2.$$

7. Let $y_1 \equiv 1$. We use the method of Section 2.4, seeking a second solution of the form $y_2 = v \cdot y_1$. Thus

$$v = \int \frac{1}{y_1}^2 e^{\int p(x)\, dx}\, dx,$$

where p is the coefficient of y' in the differential equation written in normal form. It follows that $v = -1/[2x^2]$. Thus $y_2 = -1/[2x^2]$.

Since the differential equation is homogeneous, we find that the general solution is thus

$$y = A + B\frac{1}{2x^2}.$$

8. Let $y_1 = x^2$. We use the method of Section 2.4 to seek a second solution of the form $y_2 = v \cdot y_1$, where

$$v = \int \frac{1}{y_1}^2 e^{-\int p(x)\,dx}\,dx,$$

and p is the coefficient of y' in the differential equation written in normalized form. Thus

$$v = \int \frac{1}{x^4} e^{-\ln x}\,dx = \frac{-x^4}{4}.$$

We conclude that $y_2 = -x^{-2}/4$. Since the equation is homogeneous, we see that the general solution is

$$y = Ax^2 + Bx^{-2}.$$

9. (a) The associated polynomial is $r^2 + 2r + 4$ with roots $r = -1 \pm i\sqrt{3}$. Thus the general solution to the differential equation is

$$y = Ae^{-x} \cos \sqrt{3}x + Be^{-x} \sin \sqrt{3}.$$

(b) The associated polynomial is $r^2 - 3r + 6$ with roots $r = 3/2 \pm i\sqrt{15}/2$. Thus solutions to the homogeneous equation are $y = e^{3x/2} \cos[\sqrt{15}/2]x$ and $y = e^{3x/2} \sin[\sqrt{15}/2]x$. For a particular solution, we guess $y_p = \alpha x^2 + \beta x + \gamma$. Substituting this expression into the differential equation and solving, we find that $y_p = x/6 + 1/12$ and the general solution of the given differential equation is

$$y = Ae^{3x/2} \cos[\sqrt{15}/2]x + Be^{3x/2} \sin[\sqrt{15}/2]x + \frac{x}{6} + \frac{1}{12}.$$

10. (a) The associated polynomial is $r^3 - 3r^2 + 2r$ with roots $r = 0, 1, 2$. The solutions of the homogeneous equation are $y_1 = e^0 \equiv 1$, $y_2 = e^x$, and $y_3 = e^{2x}$. We guess a particular solution of the form $y_p = \alpha x^3 + \beta x^2 + \gamma x + \delta$ and find that $y_p = [1/4]x^2 + [3/4]x$. Thus the general solution of the differential equation is

$$y = A + Be^x + Ce^{2x} + \tfrac{1}{4}x^2 + \tfrac{3}{4}x.$$

(b) The associated polynomial is $r^3 - 3r^2 + 4r - 2$ with roots $r = 1, 1 \pm i$. The general solution of the differential equation is

$$y = Ae^x + Be^x \cos x + Ce^x \sin x.$$

11. We use Kepler's Third Law. We have

$$\frac{T^2}{a^3} = \frac{4\pi^2}{GM}.$$

We must be careful to use consistent units. The gravitational constant G is given in terms of grams, centimeters, and seconds. The mass of the sun is in grams. We convert the semimajor axis to centimeters: $a = 1200 \times 10^{11}$ cm $= 1.2 \times 10^{14}$ cm. Then we calculate that

$$T = \left(\frac{4\pi^2}{GM} \cdot a^3 \right)^{1/2}$$

$$= \left(\frac{4\pi^2}{(6.637 \times 10^{-8})(2 \times 10^{33})} \cdot (1.2 \times 10^{14})^3 \right)^{1/2}$$

$$\approx [5.1393 \times 10^{17}]^{1/2} \sec$$

$$= 7.16889 \times 10^8 \sec.$$

There are 3.16×10^7 seconds in an Earth year. We divide by this number to find that the time of one orbit is

$$T \approx 22.686 \text{ Earth years.}$$

Chapter 3

1. (a) We calculate that

$$\lim_{j \to +\infty} \left| \frac{a_{j+1}}{a_j} \right| = \lim_{j \to +\infty} \frac{2^{j+1}/(j+1)!}{2^j/j!} = \lim_{j \to +\infty} \frac{2}{j+1} = 0.$$

It follows that the radius of convergence is $1/0 = +\infty$.

(b) We calculate that

$$\lim_{j \to +\infty} \left| \frac{a_{j+1}}{a_j} \right| = \lim_{j \to +\infty} \frac{2^{j+1}/3^{j+1}}{2^j/3^j} = \lim_{j \to +\infty} \frac{2}{3} = \frac{2}{3}.$$

Hence the radius of convergence is $1/(2/3) = 3/2$.

2. The power series for $\cos x$ is

$$\sum_{j=0}^{\infty}(-1)^j \frac{x^{2j+1}!}{(2j+1)!}.$$

We calculate that

$$\lim_{j\to+\infty}\left|\frac{a_{j+1}}{a_j}\right| = \lim_{j\to+\infty}\frac{1!/(2(j+1)+1)!}{1/(2j+1)!}$$

$$= \lim_{j\to+\infty}\frac{1}{(2j+2)(2j+3)} = 0.$$

It follows that the radius of convergence for the power series of the cosine function is $+\infty$.

The power series for $\sin x$ is

$$\sum_{j=0}^{\infty}(-1)^j \frac{x^{2j}!}{(2j)!}.$$

We calculate that

$$\lim_{j\to+\infty}\left|\frac{a_{j+1}}{a_j}\right| = \lim_{j\to+\infty}\frac{1!/(2(j+1))!}{1/(2j)!} = \lim_{j\to+\infty}\frac{1}{(2j+1)(2j+2)} = 0.$$

It follows that the radius of convergence for the power series of the sine function is $+\infty$.

3. With $f(x) = e^x$, we see (for $|x| \leq M$) that

$$|R_n(x)| = \left|\frac{f^{(n+1)}(\xi)}{(n+1)!}x^{n+1}\right| = \left|\frac{e^\xi}{(n+1)!}x^{n+1}\right| \leq e^M \cdot \frac{M^{n+1}}{(n+1)!} \to 0$$

as $n \to \infty$. Thus the power series for e^x converges uniformly on compact sets.

With $g(x) = \sin x$ and $w(x)$ denoting either sine or cosine, we see (for $|x| \leq M$) that

$$|R_n(x)| = \left|\frac{g^{(n+1)}(\xi)}{(n+1)!}x^{n+1}\right| = \left|\frac{w(\xi)}{(n+1)!}x^{n+1}\right| \frac{M^{n+1}}{(n+1)!} \to 0$$

as $n \to \infty$. Thus the power series for $\sin x$ converges uniformly on compact sets.

The argument for $\cos x$ is similar.

4. (a) We have $y = \sum_{j=0}^{\infty}(-1)^j x^{2j}/(2j)!$, hence

$$y'' = \sum_{j=1}^{\infty}(-1)^j 2j(2j-1)x^{2j-2}/(2j)!$$

Changing the index of summation yields

$$y'' = \sum_{k=0}^{\infty}(-1)^{k+1}(2k+2)(2k+1)x^{2k}/(2k+2)!$$

$$= -\sum_{k=0}^{\infty}(-1)^k x^{2k}/(2k)! = -y.$$

(b) We have $y = \sum_{j=0}^{\infty}(-1)^j x^{2j}/((2j)^2 \cdot (2j-2)^2 \cdots 2^2)$. Then

$$\left|\frac{a_{j+1}}{a_j}\right| = \frac{1}{(2j+2)^2} \to 0$$

as $j \to +\infty$. So the series converges for all x. We calculate that

$$xy'' + y' + xy$$

$$= x\sum_{j=1}^{\infty}(-1)^j(2j)(2j-1)\frac{x^{2j-2}}{(2j)^2 \cdot (2j-2)^2 \cdots 2^2}$$

$$+ \sum_{j=1}^{\infty}(-1)^j(2j)\frac{x^{2j-1}}{(2j)^2 \cdot (2j-2)^2 \cdots 2^2}$$

$$+ x\sum_{j=0}^{\infty}(-1)^j\frac{x^{2j}}{((2j)^2 \cdot (2j-2)^2 \cdots 2^2)}$$

$$= \sum_{j=0}^{\infty}(-1)^{j+1}(2j+2)(2j+1)\frac{x^{2j+1}}{(2j+2)^2 \cdot (2j)^2 \cdots 2^2}$$

$$+ \sum_{j=0}^{\infty}(-1)^{j+1}(2j+2)\frac{x^{2j+1}}{(2j+2)^2 \cdot (2j)^2 \cdots 2^2}$$

$$+ \sum_{j=0}^{\infty}(-1)^j\frac{x^{2j+1}}{((2j)^2 \cdot (2j-2)^2 \cdots 2^2)}$$

$$= \sum_{j=0}^{\infty} \left[(-1)^{j+1}(4j^2 + 6j + 2 + 2j + 2) \right.$$

$$\left. + (-1)^j (2j + 2)^2 \right] \frac{x^{2j+1}}{(2j+2)^2 \cdot (2j)^2 \cdots 2^2}$$

$$= \sum_{j=0}^{\infty} 0 \, x^{2j+1}$$

$$= 0.$$

5. (a) If $y = \sum_{j=0}^{\infty} a_j x^j$, then $y' = \sum_{j=1}^{\infty} j a_j x^{j-1}$. The differential equation then says

$$\sum_{j=1}^{\infty} j a_j x^{j-1} = 2x \sum_{j=0}^{\infty} a_j x^j.$$

Changing indices and combining, we find that

$$\sum_{j=1}^{\infty} [2a_{j-1} - (j+1)a_{j+1}] x^j = a_1.$$

Solving the recursion gives

$$a_1 = 0$$

$$a_2 = a_0$$

$$a_3 = 0$$

$$a_4 = \frac{1}{2} a_0$$

$$a_5 = 0$$

$$a_6 = \frac{1}{3 \cdot 2} a_0$$

and so forth. We thus find the solution

$$y = a_0 \sum_{j=0}^{\infty} \frac{x^{2j}}{j!} = a_0 e^{x^2}.$$

(b) If $y = \sum_{j=0}^{\infty} a_j x^j$, then $y' = \sum_{j=1}^{\infty} j a_j x^{j-1}$. The differential equation then says

$$\sum_{j=1}^{\infty} j a_j x^{j-1} + \sum_{j=0}^{\infty} a_j x^j = 1.$$

Changing indices and combining, we find that

$$\sum_{j=0}^{\infty} [(j+1)a_{j+1} + a_j] x^j = 1.$$

Solving the recursion gives

$$a_1 = 1 - a_0$$

$$a_2 = (-1)\frac{1}{2}(1 - a_0)$$

$$a_3 = (-1)^2 \frac{1}{3 \cdot 2}(1 - a_0)$$

$$a_4 = (-1)^3 \frac{1}{4 \cdot 3 \cdot 2}(1 - a_0)$$

and so forth. We thus find the solution

$$y = 1 + (1 - a_0) \sum_{j=0}^{\infty} \frac{(-x)^j}{j!} = 1 + (1 - a_0)e^{-x}.$$

(c) If $y = \sum_{j=0}^{\infty} a_j x^j$, then $y' = \sum_{j=1}^{\infty} j a_j x^{j-1}$. The differential equation then says

$$\sum_{j=1}^{\infty} j a_j x^{j-1} - \sum_{j=0}^{\infty} a_j x^j = 2.$$

Changing indices and combining, we find that

$$\sum_{j=0}^{\infty} [(j+1)a_{j+1} - a_j] x^j = 2.$$

Solving the recursion gives

$$a_1 = a_0 + 2$$

$$a_2 = \frac{1}{2}(a_0 + 2)$$

$$a_3 = \frac{1}{3 \cdot 2}(a_0 + 2)$$

$$a_4 = \frac{1}{4 \cdot 3 \cdot 2}(a_0 + 2)$$

and so forth. We thus find the solution

$$y = -2 + (a_0 + 2) \sum_{j=0}^{\infty} \frac{x^j}{j!} = -2 + (a_0 + 2)e^x.$$

6. (a) If $y = \sum_{j=0}^{\infty} a_j x^j$, then $y' = \sum_{j=1}^{\infty} j a_j x^{j-1}$. The differential equation then says

$$x \sum_{j=1}^{\infty} j a_j x^{j-1} = \sum_{j=0}^{\infty} a_j x^j.$$

Changing indices and combining, we find that

$$\sum_{j=1}^{\infty} [j a_j - a_j] x^j = a_0.$$

Solving the recursion gives

$$a_0 = 0$$

$$a_1 = \text{arbitrary}$$

$$a_j = 0 \text{ for all } j \geq 2.$$

We thus find the solution

$$y = a_1 x.$$

(b) If $y = \sum_{j=0}^{\infty} a_j x^j$, then $y' = \sum_{j=1}^{\infty} j a_j x^{j-1}$. The differential equation, rewritten as $xy' - y = x^3$, then says

$$x \sum_{j=1}^{\infty} j a_j x^{j-1} - \sum_{j=0}^{\infty} a_j x^j = x^3.$$

Changing indices and combining, we find that

$$-a_0 + \sum_{j=1}^{\infty}[ja_j - a_j]x^j = x^3.$$

Solving the recursion gives

$$a_0 = 0$$

$$a_1 = \text{arbitrary}$$

$$a_2 = 0$$

$$a_3 = \frac{1}{2}$$

$$a_j = 0 \text{ for } j \geq 4.$$

We thus find the solution

$$y = a_1 x + \frac{x^3}{2}.$$

7. (a) If $y = \sum_{j=0}^{\infty} a_j x^j$, then $y' = \sum_{j=1}^{\infty} ja_j x^{j-1}$ and $y'' = \sum_{j=2}^{\infty} j(j-1)a_j x^{j-2}$. The differential equation then says

$$\sum_{j=2}^{\infty} j(j-1)a_j x^{j-2} + x\sum_{j=1}^{\infty} ja_j x^{j-1} + \sum_{j=0}^{\infty} a_j x^j = 0.$$

Changing indices and combining, we find that

$$\sum_{j=1}^{\infty}[(j+2)(j+1)a_{j+2} + (j+1)a_j]x^j = -2a_2 - a_0.$$

Solving the recursion gives

$$a_2 = (-1)\frac{1}{2}a_0$$

$$a_3 = -\frac{1}{3}a_1$$

$$a_4 = (-1)^2\frac{1}{4 \cdot 2}a_0$$

$$a_5 = (-1)^2\frac{1}{5 \cdot 3}a_1$$

$$a_6 = (-1)^3\frac{1}{6 \cdot 4 \cdot 2}a_0$$

and so forth. We thus find the solution

$$y = a_0 + a_0 \sum_{j=1}^{\infty} (-1)^j \frac{x^{2j}}{2j \cdot (2j-2) \cdots 2}$$

$$+ a_1 \sum_{j=1}^{\infty} (-1)^j \frac{x^{2j-1}}{(2j-1) \cdot (2j-3) \cdots 1}.$$

(b) For the series preceded by a_0, the ratio test yields

$$\left| \frac{1/[(2j+2)(2j) \cdots 2]}{1/[2j(2j-2) \cdots 2]} \right| = \frac{1}{2j+2} \to 0.$$

Thus the radius of convergence is $+\infty$.

The calculation for the other series is similar.

8. (a) If $y = \sum_{j=0}^{\infty} a_j x^j$, then $y' = \sum_{j=1}^{\infty} j a_j x^{j-1}$ and $y'' = \sum_{j=2}^{\infty} j(j-1) a_j x^{j-2}$. The differential equation then says

$$\sum_{j=2}^{\infty} j(j-1) a_j x^{j-2} + \sum_{j=0}^{\infty} a_j x^j = x^2.$$

Changing indices and combining, we find that

$$\sum_{j=0}^{\infty} [(j+2)(j+1) a_{j+2} + a_j] x^j = x^2.$$

Solving the recursion gives

$$a_2 = (-1) \frac{1}{2 \cdot 1} a_0$$

$$a_3 = (-1) \frac{1}{3 \cdot 2} a_1$$

$$a_4 = (-1)^2 \frac{1}{4 \cdot 3 \cdot 2 \cdot 1} a_0 + \frac{1}{4 \cdot 3}$$

$$a_5 = (-1)^2 \frac{1}{5 \cdot 4 \cdot 3 \cdot 2 \cdot 1} a_1$$

$$a_6 = (-1)^3 \frac{1}{6 \cdot 5 \cdot 4 \cdot 3 \cdot 2 \cdot 1} a_0 + \frac{1}{6 \cdot 5 \cdot 4 \cdot 3}$$

and so forth. We thus find the solution

$$y = a_0 \sum_{j=0}^{\infty} (-1)^j \frac{1}{(2j)!} x^{2j} + a_1 \sum_{j=0}^{\infty} (-1)^j \frac{1}{(2j+1)!} x^{2j+1}$$

$$+ \sum_{j=2}^{\infty} \frac{1}{2j(2j-1)\cdots 3} x^{2j}.$$

(b) If $y = \sum_{j=0}^{\infty} a_j x^j$, then $y' = \sum_{j=1}^{\infty} j a_j x^{j-1}$. The differential equation then says

$$\sum_{j=2}^{\infty} j(j-1)a_j x^{j-2} + \sum_{j=1}^{\infty} j a_j x^{j-1} = -x.$$

Changing indices and combining, we find that

$$\sum_{j=0}^{\infty} [(j+2)(j+1)a_{j+2} + (j+1)a_{j+1}] x^j = -x.$$

Solving the recursion gives

$$a_2 = (-1)\frac{1}{2 \cdot 1} a_1$$

$$a_3 = (-1)^2 \frac{1}{3 \cdot 2} a_1 + (-1)\frac{1}{3 \cdot 2}$$

$$a_4 = (-1)^3 \frac{1}{4 \cdot 3 \cdot 2} a_1 + (-1)^2 \frac{1}{4 \cdot 3 \cdot 2}$$

$$a_5 = (-1)^4 \frac{1}{5 \cdot 4 \cdot 3 \cdot 2} a_1 + (-1)^3 \frac{1}{5 \cdot 4 \cdot 3 \cdot 2}$$

and so forth. We thus find the solution

$$y = a_0 + a_1 \sum_{j=2}^{\infty} (-1)^{j-1} \frac{1}{j!} x^j + \sum_{j=3}^{\infty} (-1)^j \frac{1}{j!}.$$

Chapter 4

1. We calculate that

$$a_0 = \frac{1}{\pi} \int_{-\pi}^{\pi/2} \pi \, dx = \frac{3\pi}{2},$$

$$a_j = \frac{1}{\pi} \int_{-\pi}^{\pi/2} \pi \cdot \cos jx \, dx = \left[\frac{1 + (-1)^{j-1}}{2}\right] \frac{(-1)^{[j/2]}}{j} \quad \text{for } j \geq 1,$$

$$b_j = \frac{1}{\pi} \int_{-\pi}^{\pi/2} \pi \cdot \sin jx \, dx = -\left[\frac{1 + (-1)^{j}}{2}\right] \frac{(-1)^{[j/2]}}{j} \quad \text{for } j \geq 1.$$

Thus the Fourier series is

$$f(x) = \frac{3\pi}{4} + \sum_{j=1}^{\infty} \left[\frac{1 + (-1)^{j-1}}{2}\right] \frac{(-1)^{[j/2]}}{j} \cos jx$$

$$+ \sum_{j=1}^{\infty} -\left[\frac{1 + (-1)^{j}}{2}\right] \frac{(-1)^{[j/2]}}{j} \sin jx.$$

2. We calculate that

$$a_0 = \frac{1}{\pi} \int_{0}^{\pi/2} 1 \, dx = \frac{1}{2},$$

$$a_j = \frac{1}{\pi} \int_{0}^{\pi/2} 1 \cdot \cos jx \, dx = \left[\frac{1 + (-1)^{j-1}}{2}\right] \frac{(-1)^{[j/2]}}{j\pi} \quad \text{for } j \geq 1,$$

$$b_j = \frac{1}{\pi} \int_{0}^{\pi/2} 1 \cdot \sin jx \, dx = -\left[\frac{1 + (-1)^{j}}{2}\right] \frac{(-1)^{[j/2]}}{j\pi} \quad \text{for } j \geq 1.$$

Thus the Fourier series is

$$f(x) = \frac{1}{4} + \sum_{j=1}^{\infty} \left[\frac{1 + (-1)^{j-1}}{2}\right] \frac{(-1)^{[j/2]}}{j\pi} \cos jx$$

$$+ \sum_{j=1}^{\infty} -\left[\frac{1 + (-1)^{j}}{2}\right] \frac{(-1)^{[j/2]}}{j\pi} \sin jx.$$

3. We calculate that

$$a_0 = \frac{1}{\pi} \int_0^\pi \sin x \, dx = \frac{2}{\pi},$$

$$a_j = \frac{1}{\pi} \int_0^\pi \sin x \cos jx \, dx = \frac{1}{j^2+1} \frac{1}{\pi} ((-1)^j + 1) \quad \text{for } j \geq 1,$$

$$b_j = \frac{1}{\pi} \int_0^\pi \sin x \cos jx \, dx = 0.$$

Thus the Fourier series is

$$f(x) = \frac{1}{\pi} + \sum_{j=1}^\infty \frac{1}{j^2+1} \frac{1}{\pi} ((-1)^j + 1) \cos jx.$$

4. We calculate that

$$a_0 = \frac{1}{\pi} \int_{-\pi}^0 -\pi \, dx + \frac{1}{\pi} \int_0^\pi x \, dx = -\pi + \frac{\pi}{2} = -\frac{\pi}{2},$$

$$a_j = \frac{1}{\pi} \int_{-\pi}^0 -\pi \cos jx \, dx + \frac{1}{\pi} \int_0^\pi x \cos jx \, dx = \frac{(-1)^j - 1}{\pi j^2} \quad \text{for } j \geq 1,$$

$$b_j = \frac{1}{\pi} \int_{-\pi}^0 -\pi \sin jx \, dx + \frac{1}{\pi} \int_0^\pi x \sin x \, dx = \frac{1 + 2(-1)^{j+1}}{j}.$$

Thus the Fourier series is

$$f(x) = -\frac{\pi}{4} + \sum_{j=1}^\infty \frac{(-1)^j - 1}{\pi j^2} \cos jx + \sum_{j=1}^\infty \frac{1 + 2(-1)^{j+1}}{j} \sin jx.$$

5. We calculate that

$$a_0 = \frac{1}{\pi} \int_0^\pi x^2 \, dx = \frac{\pi^2}{3},$$

$$a_j = \frac{1}{\pi} \int_0^\pi x^2 \cos jx \, dx = \frac{2(-1)^j}{j^2} \quad \text{for } j \geq 1,$$

$$b_j = \frac{1}{\pi} \int_0^\pi x^2 \sin jx \, dx = \frac{2}{\pi j^3} [(-1)^j - 1] - \frac{\pi(-1)^j}{j}.$$

Thus the Fourier series is

$$f(x) = \frac{\pi^2}{6} + 2\sum_{j=1}^{\infty}(-1)^j \frac{\cos jx}{j^2}$$

$$+ \pi \sum_{j=1}^{\infty}(-1)^{j+1}\frac{\sin jx}{j} - \frac{4}{\pi}\sum_{j=1}^{\infty}\frac{\sin(2j-1)x}{(2j-1)^3}.$$

6. We calculate that $(-x)^5 \sin(-x) = x^5 \sin x$, so this function is even.
   We calculate that $e^{-x} \neq e^x$, $e^{-x} \neq -e^x$, so this function is neither even nor odd.
   We calculate that $(\sin(-x))^3 = -(\sin x)^3$, so this function is odd.
   We calculate that $\sin(-x)^2 = \sin x^2$, so this function is even.
   We calculate that $(-x) + (-x)^2 + (-x)^3 \neq x + x^2 + x^3$, $(-x) + (-x)^2 + (-x)^3 \neq -(x + x^2 + x^3)$, so this function is neither even nor odd.
   We calculate that $\ln\dfrac{1+(-x)}{1-(-x)} \neq \ln\dfrac{1+x}{1-x}$, $\ln\dfrac{1+(-x)}{1-(-x)} \neq -\ln\dfrac{1+x}{1-x}$, so this function is neither even nor odd.

7. We write

$$f(x) = \frac{1}{2}\left[\frac{f(x)+f(-x)}{2}\right] + \frac{1}{2}\left[\frac{f(x)-f(-x)}{2}\right] \equiv f_e(x) + f_o(x).$$

Then

$$f_e(-x) = \frac{1}{2}\left[\frac{f(-x)+f(x)}{2}\right] = f_e(x),$$

so that f_e is even. Also

$$f_o(-x) = \frac{1}{2}\left[\frac{f(-x)-f(-(-x))}{2}\right] = -\frac{1}{2}\left[\frac{f(x)-f(-x)}{2}\right] = -f_o(x),$$

so that f_o is odd.

8. We calculate the sine coefficients of $\widetilde{f}$:

$$b_j = \frac{2}{\pi}\int_0^{\pi}\frac{\pi}{4}\sin jx\,dx = \frac{1}{2}\frac{(-\cos jx)}{j}\bigg|_0^{\pi} = \frac{1}{2}\left[\frac{(-1)^{j+1}+1}{2}\right].$$

Thus the even coefficients vanish and the odd $(2j+1)$th coefficients are $1/(2j+1)$. We see then that

$$\frac{\pi}{4} = \sin x + \frac{\sin 3x}{3} + \frac{\sin 5x}{5} + \cdots.$$

Evaluating this series expansion at $x = \pi/2$ gives a famous series representation for π:

$$\frac{\pi}{4} = 1 - \frac{1}{3} + \frac{1}{5} - + \cdots .$$

The cosine coefficients of $\tilde{\tilde{f}}$ are

$$a_0 = \frac{\pi}{4}$$

and, for $j \geq 1$,

$$a_j = \frac{2}{\pi} \int_0^\pi \frac{\pi}{4} \sin jx \, dx = \frac{1}{2} \frac{\sin jx}{j} \Big|_0^\pi = 0.$$

Thus the cosine expansion of f is

$$f(x) = \frac{\pi}{4}.$$

9. We calculate that

$$b_j = \frac{2}{\pi} \int_0^\pi \sin x \sin jx \, dx$$

$$= \frac{2}{\pi} (-\cos x) \sin jx \Big|_0^\pi + \frac{2}{\pi} \int_0^\pi \cos x (j \cos jx) \, dx$$

$$= \frac{2}{\pi} \sin x (j \cos jx) \Big|_0^\pi - \frac{2}{\pi} \int_0^\pi \sin x (-j^2 \sin jx) \, dx.$$

It follows that $b_j = 0$ for $j \geq 2$. Also we calculate easily that $b_1 = 1$. Thus the sine series expansion of $f(x) = \sin x$ is

$$f(x) = \sin x.$$

A similar calculation shows that the cosine series expansion of $f(x) = \sin x$ is

$$f(x) = \cos x = \sum_{j=2}^\infty \frac{1}{1 - j^2} \frac{2}{\pi} [(-1)^j + 1] \cos jx.$$

10. (a) We calculate that

$$a_0 = \int_{-1}^0 (1 + x) \, dx + \int_0^1 (1 - x) \, dx = 1$$

and

$$a_j = \int_{-1}^{0} (1+x)\cos(j\pi x)\,dx + \int_{0}^{1} (1-x)\cos(j\pi x)\,dx$$

$$= \frac{\sin(j\pi x)}{n\pi}(1+x)\Big|_{-1}^{0} - \int_{-1}^{0} \frac{\sin(j\pi x)}{n\pi}\,dx$$

$$+ \frac{\sin(j\pi x)}{j\pi}(1-x)\Big|_{0}^{1} - \int_{0}^{1} \frac{\sin(j\pi x)}{n\pi}(-1)\,dx$$

$$= \frac{2}{j^2\pi^2}\Big[1+(-1)^j\Big].$$

A similar calculation shows that

$$b_j = 0 \qquad \text{for all } j.$$

As a result,

$$f(x) = 1 + \sum_{j=1}^{\infty} \frac{2}{j^2\pi^2}\Big[1+(-1)^j\Big]\cos jx.$$

(b) We calculate that

$$a_0 = \frac{1}{2}\int_{-2}^{0} -x\,dx + \frac{1}{2}\int_{0}^{2} x\,dx = 1$$

and

$$a_j = \frac{1}{2}\int_{-2}^{0} (-x)\cos jx\,dx + \frac{1}{2}\int_{0}^{2} x\cos jx\,dx$$

$$= \int_{0}^{2} x\cos jx\,dx$$

$$= \frac{\sin(j\pi x/2)}{j\pi/2}\cdot x\Big|_{0}^{2} - \int_{0}^{2} \frac{\sin(j\pi x/2)}{j\pi/2}\,dx$$

$$= \frac{2}{j^2\pi^2}[(-1)^j - 1].$$

By the oddness of $|x|\sin jx$, we see that $b_j = 0$ for all j. As a result,

$$f(x) = 1 + \sum_{j=1}^{\infty} \frac{2}{j^2\pi^2}[(-1)^j - 1]\cos jx.$$

Chapter 5

1. (a) Since $y(0) = 0$, the only relevant solutions to the differential
    equation are $y_\lambda(x) = \sin \lambda x$. Since $y(\pi/2) = 0$, we find that
    $\lambda = 4, 16, 36, \ldots, (2n)^2, \ldots$. The corresponding eigenfunctions are
    $\sin 2x, \sin 4x, \sin 6x$, etc.

    (b) Since $y(0) = 0$, the only relevant solutions to the differential
    equation are $y_\lambda(x) = \sin \lambda x$. Since $y(2\pi) = 0$, we find that
    $\lambda = 1/4, 1, 9/4, \ldots, n^2/4, \ldots$. The corresponding eigenfunctions
    are $\sin(1/2)x, \sin x, \sin(3/2)x$, etc.

    (c) Since $y(0) = 0$, the only relevant solutions to the differential equa-
    tion are $y_\lambda(x) = \sin \lambda x$. Since $y(1) = 0$, we find that $\lambda = \pi^2, 4\pi^2,$
    $9\pi^2, \ldots, n^2\pi^2, \ldots$. The corresponding eigenfunctions are $\sin \pi x$,
    $\sin 2\pi x, \sin 3\pi x$, etc.

2. (a) We need the sine series expansion of f:

    $$b_j = \frac{2}{\pi} \int_0^{\pi/2} \frac{2x}{\pi} \sin jx \, dx + \frac{2}{\pi} \int_{\pi/2}^{\pi} \frac{2(\pi - x)}{\pi} \, dx$$

    $$= \frac{4}{\pi^2} \frac{-\cos jx}{j} \cdot x \Big|_0^{\pi/2} + \frac{4}{\pi^2} \int_0^{\pi/2} \frac{\cos jx}{j} \, dx$$

    $$+ \frac{4}{\pi^2} \frac{-\cos jx}{j} \cdot (\pi - x) \Big|_{\pi/2}^{\pi} + \frac{4}{\pi^2} \int_{\pi/2}^{\pi} \frac{\cos jx}{j} \cdot (-1) \, dx$$

    $$= \frac{4}{\pi^2 j^2} [(-1)^{j+1} + 1] \cdot (-1)^{\lfloor j/2 \rfloor}.$$

    It follows that the solution of the vibrating string for this data is

    $$y(x, t) = \sum_{j=1}^{\infty} \left[\frac{4}{\pi^2 j^2} [(-1)^{j+1} + 1] \cdot (-1)^{\lfloor j/2 \rfloor} \right] \sin jx \cos jt.$$

    (b) We need the sine series expansion of f:

    $$b_j = \frac{2}{\pi} \int_0^{\pi} \frac{1}{\pi} x(\pi - x) \sin jx \, dx$$

    $$= \frac{2}{\pi^2} \left(-\frac{\cos jx}{j} (x\pi - x^2) \right) \Big|_0^{\pi} + \frac{2}{\pi^2} \int_0^{\pi} \frac{\cos jx}{j} (\pi - 2x) \, dx$$

    $$= \frac{4}{\pi^2} \left[\frac{1}{j^3} - \frac{(-1)^j}{j^3} \right].$$

It follows that the solution of the vibrating string for this data is

$$y(x, t) = \sum_{j=1}^{\infty} \left(\frac{4}{\pi^2} \left[\frac{1}{j^3} - \frac{(-1)^j}{j^3} \right] \right) \sin jx \cos jt.$$

3. It is easy to calculate that

$$b_1 = c \sin x$$

and all other b's are equal to zero. Thus the solution of the vibrating string with this initial data is

$$y(x, t) = c \cdot \sin x \cos t.$$

We see that, for fixed time $t = t_0$, the curve has the shape

$$y(x, t_0) = [c \cos t_0] \sin x.$$

Plainly this is a standard sine curve with modified amplitude.

4. We pose a solution of the form $y(x, t) = \alpha(x)\beta(t)$. Plugging this into the differential equation $y_{xx} = y_{tt}$ (we take $a = 1$ for simplicity), we find that

$$\alpha''(x)\beta(t) = \alpha(x)\beta''(t).$$

This leads to

$$\frac{\alpha''(x)}{\alpha(x)} = \frac{\beta''(t)}{\beta(t)}.$$

Thus we have the differential equations

$$\alpha''(x) - \mu\alpha(x) = 0 \qquad\qquad (a)$$

and

$$\beta''(t) - \mu\beta(t) = 0. \qquad\qquad (b)$$

Of course the boundary conditions tell us that $\mu = -\lambda$, where $\lambda > 0$. And, as usual, the eigenfunctions for equation (a) are $\sin \sqrt{\lambda}x$. Thus

$$y(x, t) = \sin \sqrt{\lambda}x[A \cos \sqrt{\lambda}t + B \sin \sqrt{\lambda}t].$$

But now the fact that $u(x, 0) = 0$ tells us that $A = 0$. Hence

$$u(x, t) = B \sin \sqrt{\lambda}x \sin \sqrt{\lambda}t.$$

As usual, $\lambda = j^2$ for $j = 1, 2, \ldots$. So the general solution is a linear combination of terms $\sin \sqrt{\lambda}x \sin \sqrt{\lambda}t$.

5. Let b_j be the coefficients of the sine series expansion of the function $f(x)$. Then the solution of the heat equation will be

$$y(x, t) = \sum_{j=1}^{\infty} b_j e^{-j^2 t} \sin jx.$$

6. (a) We calculate the Fourier series for the function f:

$$a_j = \frac{1}{\pi} \int_{-\pi}^{\pi} \cos \frac{\theta}{2} \cos j\theta \, d\theta$$

$$= \frac{1}{\pi} \int_{-\pi/2}^{\pi/2} (\cos \psi)(\cos 2j\psi) 2 \, d\psi$$

$$= \frac{2}{\pi} \sin \psi \cos 2j\psi \bigg|_{-\pi/2}^{\pi/2} - \frac{2}{\pi} \int_{-\pi/2}^{\pi/2} \sin \psi (-2j \sin 2j\psi) \, d\psi$$

$$= \frac{4}{\pi} (-1)^j + \frac{4j^2}{\pi} \int_{-\pi/2}^{\pi/2} 2 \cos \psi \cos 2j\psi \, d\psi$$

$$= \frac{4(-1)^j}{\pi(1 - 4j^2)}.$$

A similar calculation shows that

$$b_j = 0 \qquad \text{for all } j.$$

Thus the solution to the Dirichlet problem is

$$w(r, \theta) = \frac{2}{\pi} + \sum_{j=1}^{\infty} r^j \frac{4(-1)^j}{\pi(1 - 4j^2)} \cos j\theta.$$

(b) We calculate the Fourier series for the function f:

$$a_j = \frac{1}{\pi} \int_{-\pi}^{\pi} \theta \cos j\theta \, d\theta$$

$$= \frac{1}{\pi j} (\sin j\theta) \cdot \theta \bigg|_{-\pi}^{\pi} - \frac{1}{\pi j} \int_{-\pi}^{\pi} \sin j\theta \, d\theta$$

$$= 0.$$

Similarly,

$$b_j = \frac{2(-1)^{j+1}}{j}.$$

Thus the solution to the Dirichlet problem is

$$w(r, \theta) = \sum_{j=1}^{\infty} r^j \frac{2(-1)^{j+1}}{j} \sin j\theta.$$

(c) We calculate the Fourier series for the function f:

$$a_1 = \frac{1}{\pi} \int_0^{\pi} \sin \theta \cos \theta \, d\theta = 0,$$

$$a_j = \frac{1}{\pi} \int_0^{\pi} \sin \theta \cos j\theta \, d\theta$$

$$= -\frac{\cos \theta \cos j\theta}{\pi} \bigg|_0^{\pi} + \frac{1}{\pi} \int_0^{\pi} \cos \theta(-j \sin j\theta) \, d\theta$$

$$= j^2 \frac{1}{\pi} \int_0^{\pi} \sin \theta \cos j\theta \, d\theta,$$

hence

$$a_j = \frac{1}{1 - j^2} \left(\frac{1 + (-1)^j}{\pi} \right) \quad \text{for } j \geq 1 \quad \text{for } j \neq 1,$$

and a similar calculation shows that

$$b_j = 0 \qquad \text{for all } j.$$

Thus the solution to the Dirichlet problem is

$$w(r, \theta) = \frac{2}{\pi} + \sum_{j=2}^{\infty} r^j \frac{1}{1 - j^2} \left(\frac{1 + (-1)^j}{2} \right) \cos j\theta.$$

7. Using polar coordinates, if the point (r, θ) lies in the disc $D(0, R)$ with $0 \leq r < R$, then the point $(r/R, \theta)$ lies in $D(0, 1)$ with $0 \leq r/R < 1$. The process can be reversed as well. Thus if $f(\theta)$ is a boundary function for $D(0, R)$ with coefficients a_j, b_j of its Fourier series, then

$$(r, \theta) \mapsto \left(\frac{r}{R}, \theta \right) \mapsto \frac{a_0}{2} + \sum_{j=1}^{\infty} \left(\frac{r}{R} \right)^j [a_j \cos j\theta + b_j \sin j\theta]$$

solves the Dirichlet problem on $D(0, R)$.

8. We calculate the sine series of f:

$$b_j = \frac{2}{\pi} \int_0^{\pi/2} x \sin jx \, dx + \frac{2}{\pi} \int_{\pi/2}^{\pi} (\pi - x) \sin x \, dx$$

$$= \frac{2}{\pi} \left(\frac{-\cos jx}{j} \right) x \Big|_0^{\pi/2} + \frac{2}{\pi} \int_0^{\pi/2} \frac{\cos jx}{j} \, dx$$

$$+ \frac{2}{\pi} \left(\frac{-\cos jx}{j} \right) (\pi - x) \Big|_{\pi/2}^{\pi} + \frac{2}{\pi} \int_{\pi/2}^{\pi} \frac{\cos jx}{j} \cdot (-1) \, dx$$

$$= \frac{2}{j^2 \pi} \left[(-1)^{j+1} + 1 \right] (-1)^{\lfloor j/2 \rfloor}.$$

It follows that the solution of the vibrating string for this data is

$$y(x, t) = \sum_{j=1}^{\infty} \frac{2}{j^2 \pi} \left[(-1)^{j+1} + 1 \right] (-1)^{\lfloor j/2 \rfloor} \sin jx \cos jt.$$

9. We calculate the Fourier series of f:

$$a_0 = \frac{1}{\pi} \int_{-\pi}^0 2\theta \, d\theta = -\pi,$$

$$a_j = \frac{1}{\pi} \int_{-\pi}^0 2\theta \cos j\theta \, d\theta$$

$$= \frac{1}{\pi} \frac{\sin j\theta}{j} \cdot 2\theta \Big|_{-\pi}^0 - \frac{1}{\pi} \int_{-\pi}^0 \frac{\sin j\theta}{j} \cdot 2 \, d\theta$$

$$= \frac{2}{j^2 \pi} [1 + (-1)^{j+1}],$$

$$b_j = \frac{1}{\pi} \int_{-\pi}^0 2\theta \sin j\theta \, d\theta$$

$$= \frac{1}{\pi} \frac{-\cos j\theta}{j} \cdot 2\theta \Big|_{-\pi}^0 + \frac{1}{\pi} \int_{-\pi}^0 \frac{\cos j\theta}{j} \cdot 2 \, d\theta$$

$$= \frac{2(-1)^{j+1}}{j}.$$

As a result, the solution of the Dirichlet problem with this data is

$$w(r, \theta) = -\pi + \sum_{j=1}^{\infty} r^j \left[\frac{2}{j^2 \pi} [1 + (-1)^{j+1}] \cos j\theta + \frac{2(-1)^{j+1}}{j} \sin j\theta \right].$$

Chapter 6

1. We calculate that

$$L[x^n](p) = \int_0^\infty x^n e^{-px}\, dx$$

$$= \frac{e^{-px}}{-p} x^n \bigg|_0^\infty + \int_0^\infty \frac{e^{-px}}{p} \cdot n x^{n-1}\, dx$$

$$= \cdots$$

$$= \frac{n!}{p^n} \int_0^\infty e^{-px}\, dx$$

$$= \frac{n!}{p^{n+1}};$$

$$L[e^{ax}](p) = \int_0^\infty e^{ax} e^{-px}\, dx = \int_0^\infty e^{(a-p)x}\, dx = \frac{1}{p-a};$$

$$L[\sin ax](p) = \int_0^\infty \sin ax\, e^{-px}\, dx$$

$$= \frac{e^{-px}}{-p} \sin ax \bigg|_0^\infty + \int_0^\infty \frac{e^{-px}}{p} (a \cos ax)\, dx$$

$$= \frac{a}{p} \int_0^\infty \cos ax\, e^{-px}\, dx$$

$$= \frac{a}{p} \frac{e^{-px}}{-p} \cos ax \bigg|_0^\infty + \frac{a}{p} \int_0^\infty \frac{e^{-px}}{p} (-a \sin ax)\, dx$$

$$= \frac{a}{p^2} - \frac{a^2}{p^2} L[\sin ax](p).$$

It follows then that

$$L[\sin ax](p) = \frac{a}{p^2 + a^2}.$$

2. (a) We calculate that

$$L[\sinh ax](p) = L\left[\frac{e^a - e^{-ax}}{2}\right]$$

$$= \frac{1}{2}\left[L[e^{ax}](p) - L[e^{-ax}](p)\right]$$

$$= \frac{1}{2}\left[\frac{1}{p-a} - \frac{1}{p+a}\right]$$

$$= \frac{a}{p^2 - a^2}.$$

(b) We calculate that

$$L[\cosh ax](p) = L\left[\frac{e^a + e^{-ax}}{2}\right]$$

$$= \frac{1}{2}\left[L[e^{ax}](p) + L[e^{-ax}](p)\right]$$

$$= \frac{1}{2}\left[\frac{1}{p-a} + \frac{1}{p+a}\right]$$

$$= \frac{p}{p^2 - a^2}.$$

3. (a) $L[10](p) = 10L[1](p) = \dfrac{10}{p}$

(b) $L[x^5 + \cos 2x](p) = L[x^5](p) + L[\cos 2x](p) = \dfrac{5!}{p^6} + \dfrac{p}{p^2 + 4}$

(c) $L[2e^{3x} - 4\sin 5x](p) = 2L[e^{3x}](p) - 4L[\sin 5x](p) = \dfrac{2}{p-3} -$

$\dfrac{20}{p^2 + 25}$

(d) $L[4\sin x \cos x + 2e^{-x}](p) = 2L[2\sin 2x](p) + 2L[e^{-x}](p) = \dfrac{4}{p^2 + 4} + \dfrac{2}{p+1}$

(e) $L[x^6 \sin^2 3x + x^6 \cos^2 3x](p) = L[x^6] = \dfrac{6!}{p^7}.$

4. (a) $L[x^5 e^{-2x}](p) = L[e^{-2x}x^5](p) = L[x^5](p+2) = \dfrac{5!}{(p+2)^6}$

(b) $L[(1 - x^2)e^{-x}](p) = L[e^{-x}](p) - L[e^{-x}x^2](p) = \dfrac{1}{p+1} -$

$L[x^2](p+1) = \dfrac{1}{p+1} - \dfrac{2!}{(p+1)^3}$

(c) $L[e^{3x}\cos 2x](p) = L[\cos 2x](p - 3) = \dfrac{p}{(p-3)^2 + 4}$

5. (a) $L^{-1}\left[\dfrac{6}{p^2 + 9}\right](x) = 2\sin 3x$, hence $L^{-1}\left[\dfrac{6}{(p+2)^2 + 9}\right] = e^{-2x}\sin 3x$

(b) $L^{-1}[3!/p^4](x) = x^3$, hence $L^{-1}[12/p^4](x) = 2x^3$ and $L^{-1}[12/(p+3)^4](x) = 2e^{-3x}x^3$

(c) $L^{-1}\left[\dfrac{p+3}{p^2 + 2p + 5}\right](x) = L^{-1}\left[\dfrac{p+3}{(p+1)^2 + 4}\right] =$

$L^{-1}\left[\dfrac{p+1}{(p+1)^2 + 4}\right] + L^{-1}\left[\dfrac{2}{(p+1)^2 + 4}\right].$ As a result,

$L^{-1}\left[\dfrac{p+3}{p^2 + 2p + 5}\right](x) = e^{-x}\cos 2x + \sin 2x$

6. (a) The Laplace transform of the differential equation is

$$pY - y(0) + Y = \frac{1}{p - 2},$$

hence

$$pY + Y = \frac{1}{p - 2}.$$

We find that

$$Y = \frac{1}{(p - 2)(p + 1)} = \frac{1/3}{p - 2} + \frac{-1/3}{p + 1}.$$

It follows that

$$y = \tfrac{1}{3}e^{2x} - \tfrac{1}{3}e^{-x}.$$

(b) The Laplace transform of the differential equation is

$$p^2 Y - py(0) - y'(0) - 4(pY - y(0)) + 4Y = 0,$$

hence

$$Y \cdot (p^2 - 4p + 4) = 3$$

or

$$Y = 3 \cdot \frac{1}{(p-2)^2}.$$

It follows that

$$y = 3xe^{2x}.$$

(c) The Laplace transform of the differential equation is

$$p^2Y - py(0) - y'(0) + 2[pY - y(0)] + 2Y = 2,$$

hence

$$Y = \frac{3}{p^2 + 2p + 2} = \frac{1}{p} - \frac{p+1}{(p+1)^2 + 1}.$$

It follows that

$$y = 1 - e^{-x}\cos x.$$

7. The Laplace transform of the integral equation is

$$pY - y(0) + 4Y + 5\frac{Y}{p} = \frac{1}{p+1},$$

hence

$$Y\left(p + 4 + \frac{5}{p}\right) = \frac{1}{p+1}$$

or

$$Y = \frac{1}{2}\frac{p+2}{(p+2)^2 + 1} + \frac{3/2}{(p+2)^2 + 1} + \frac{-1/2}{p+1}.$$

It follows that

$$y = \frac{1}{2}e^{-2x}\cos x + \frac{3}{2}e^{-2x}\sin x - \frac{1}{2}e^{-x}.$$

8. (a) We calculate that

$$L[x^2 \sin ax](p) = -\frac{d}{dp} L[\sin ax](p)$$

$$= \frac{d^2}{dp^2} L[\sin ax](p)$$

$$= \frac{d^2}{dp^2} \left(\frac{a}{p^2 + a^2} \right)$$

$$= \frac{6ap^2 - 2a^3}{(p^2 + a^2)^3}.$$

(b) We calculate that

$$L[xe^x](p) = -\frac{d}{dp} L[e^x](p)$$

$$= -\frac{d}{dp} \left(\frac{1}{p-1} \right)$$

$$= \frac{1}{(p-1)^2}.$$

9. (a) Taking the Laplace transform, we find that

$$Y = \frac{1}{p} - L[x](p) \cdot Y = \frac{1}{p} - \frac{1}{p^2} Y,$$

hence

$$Y = \frac{p}{p^2 + 1}.$$

We conclude that

$$y = \cos x.$$

(b) We calculate the Laplace transform:

$$Y = L\left[1 + \int_0^x e^{-t} y(t) \, dt \right] (p-1)$$

$$= \frac{1}{p-1} + \frac{L[e^{-t} y(t)](p-1)}{p-1}$$

$$= \frac{1}{p-1} + \frac{Y(p)}{p-1}.$$

It follows that

$$Y = \frac{1}{p-2}$$

so that

$$y = e^{2x}.$$

10. (a) The convolution is

$$\int_0^x \sin at\, dt = -\frac{\cos at}{a}\bigg|_0^x = -\frac{\cos ax}{a} + \frac{1}{a}.$$

(b) The convolution is

$$\int_0^x e^{a(x-t)} e^{bt}\, dt = e^{ax} \int_0^x e^{(b-a)t}\, dt = \frac{e^{bx}}{b-a} - \frac{e^{ax}}{b-a}.$$

11. Taking the Laplace transform, we find that

$$(p^2 Y - pA - B) - 5(pY - A) + 4Y = 0,$$

hence

$$Y = \frac{pA - 5A + B}{p^2 - 5p + 4}.$$

As a result,

$$Y = \frac{1}{3} \cdot \frac{pA + (-5A + B)}{p - 4} - \frac{1}{3} \cdot \frac{pA + (-5A + B)}{p - 1}$$

$$= \frac{-A + B}{3} \cdot \frac{1}{p - 4} - \frac{-4A + B}{3} \cdot \frac{1}{p - 1}.$$

It follows that

$$y = \frac{-A + B}{3} e^{4x} - \frac{-4A + B}{3} e^{x}.$$

12. We write

$$g(t) = u(t - 3) \cdot (t - 1),$$

where u is the unit step function. Then

$$L[g](p) = L[tu(t-3)](p) - L[u(t-3)](p)$$

$$= -\frac{d}{dp}L[u(t-3)](p) - L[u(t-3)](p)$$

$$= -\frac{d}{dp}\left(e^{-3p}L[u](p)\right) - e^{-3p}L[u](p)$$

$$= 2e^{-3p}L[u](p) + e^{-3p}L[tu(t)](p)$$

$$= 2e^{-3p}L[u](p) + e^{-3p}L[t](p)$$

$$= 2e^{-3p}L[u](p) + e^{-3p}\frac{1}{p^2}$$

$$= e^{-3p}\left(\frac{2p+1}{p^2}\right).$$

Chapter 7

1. (a) Of course $x_0 = 0$, $y_0 = 1$. We calculate that

$$y_1 = 1 + 0.1(2 \cdot 0 + 2 \cdot 1) = 1.2,$$

$$y_2 = 1.2 + 0.1(2 \cdot 0.1 + 2 \cdot 1.2) = 1.46,$$

$$y_3 = 1.46 + 0.1(2 \cdot 0.2 + 2. \cdot 1.46) = 1.792.$$

Thus 1.792 is our Euler approximation to $y(0.3)$.

The initial value problem $y' - 2y = 2x$, $y(0) = 1$, may be solved explicitly with solution $y = -x - 1/2 + [3/2]e^{2x}$. We see then that the exact value of $y(0.3)$, to three decimal places, is $y(0.3) \approx 1.933$. The approximation is off by about 8 percent.

(b) Of course $x_0 = 0$, $y_0 = 1$. We calculate that

$$y_1 = 1 + 0.1 \cdot \frac{1}{1} = 1.1,$$

$$y_2 = 1.1 + 0.1 \cdot \frac{1}{1.1} = 1.1909,$$

$$y_3 = 1.1909 + 0.1 \cdot \frac{1}{1.1909} = 1.275.$$

Thus 1.275 is our Euler approximation to $y(0.3)$.

The initial value problem $y' = 1/y$, $y(0) = 1$, may be solved explicitly with solution $y = \sqrt{2x + 1}$. We see that the exact value of $y(0.3)$, to three decimal places, is $y(0.3) \approx 1.265$. The approximation is off by about 0.8 percent.

2. (a) From Exercise 1(a) we know that the explicit solution of the initial value problem is

$$y = -x - \frac{1}{2} + \frac{3}{2}e^{2x}.$$

It follows that

$$y'' = 6e^{2x}$$

and

$$|y''| \le 6e^2.$$

Then

$$|\epsilon_j| \le \frac{6e^2 h^2}{2} = 3e^2 h^2.$$

Finally, the total discretization error is

$$|E_n| \le \frac{3e^2 h^2}{h} = 3e^2 h = 3e^2 \cdot 0.2 \approx 4.433.$$

(b) From Exercise 1(b) we know that the explicit solution of the initial value problem is

$$y = \sqrt{2x + 1}.$$

It follows that

$$y'' = -(2x + 1)^{-3/2}$$

and

$$|y''| \le 1.$$

Then

$$|\epsilon_j| \le \frac{1 \cdot h^2}{2}.$$

Finally, the total discretization error is

$$|E_n| \le \frac{h^2/2}{h} = \frac{h}{2} = 0.1.$$

3. (a) Of course $x_0 = 0$, $y_0 = 1$. We calculate that $z_1 = 1.2$ and

$$y_1 = 1 + \frac{0.1}{2}[(2 \cdot 0 + 2 \cdot 1) + (2 \cdot 0.1 + 2 \cdot 1.2)] = 1.23.$$

Also $z_2 = 1.496$ and

$$y_2 = 1.23 + \frac{0.1}{2}[(2 \cdot 0.1 + 2 \cdot 1.23) + (2 \cdot 0.2 + 2 \cdot 1.496)] = 1.533.$$

Finally, $z_3 = 1.879$ and

$$y_3 = 1.533 + \frac{0.1}{2}[(2 \cdot 0.2 + 2 \cdot 1.533) + (2 \cdot 0.3 + 2 \cdot 1.879)] = 1.924.$$

Thus 1.924 is our improved Euler approximation to $y(0.3)$.

We see from Exercise 1(a) that the exact value, accurate to three decimal places, of $y(0.3)$ is $y(0.3) = 1.933$. The approximation is off by about 0.5 percent.

(b) Of course $x_0 = 0$, $y_0 = 1$. We calculate that $z_1 = 1.1$ and

$$y_1 = 1 + \frac{0.1}{2}\left[\frac{1}{1} + \frac{1}{1.1}\right] = 1.095.$$

Also $z_2 = 1.186$ and

$$y_2 = 1.095 + \frac{0.1}{2}\left[\frac{1}{1.095} + \frac{1}{1.186}\right] = 1.183.$$

Finally, $z_3 = 1.268$ and

$$y_3 = 1.183 + \frac{0.1}{2}\left[\frac{1}{1.183} + \frac{1}{1.268}\right] = 1.265.$$

Thus 1.265 is our improved Euler approximation to $y(0.3)$.

We see from Exercise 1(b) that the exact value, accurate to three decimal places, of $y(0.3)$ is $y(0.3) = 1.265$. The approximation agrees with the exact answer to three decimal places.

4. (a) Of course $x_0 = 0$, $y_0 = 1$. We calculate that

$$m_1 = (0.1) \cdot f(0, 1) = (0.1) \cdot (2 \cdot 0 + 2 \cdot 1) = 0.2,$$

$$m_2 = (0.1) \cdot f(0.05, 1.1) = 0.23,$$

$$m_3 = (0.1) \cdot f(0.05, 1.115) = 0.233.$$

$$m_4 = (0.1) \cdot f(0.1, 1.233) = 0.267.$$

Thus

$$y_1 = 1 + \tfrac{1}{6}(0.2 + 0.46 + 0.466 + 0.267) = 1.232.$$

Now $x_1 = 0.1$, $y_1 = 1.232$. We calculate that

$$m_1 = (0.1) \cdot f(0.1, 1.232) = 0.266,$$

$$m_2 = (0.1) \cdot f(0.15, 1.365) = 0.303,$$

$$m_3 = (0.1) \cdot f(0.15, 1.384) = 0.3068,$$

$$m_4 = (0.1) \cdot f(0.2, 1.5388) = 0.348.$$

Thus

$$y_2 = 1.232 + \tfrac{1}{6}(0.266 + 0.606 + 0.6136 + 0.348) = 1.5376.$$

Now $x_2 = 0.2$, $y_2 = 1.5376$. We calculate that

$$m_1 = (0.1) \cdot f(0.2, 1.5376) = 0.3476,$$

$$m_2 = (0.1) \cdot f(0.25, 1.712) = 0.3924,$$

$$m_3 = (0.1) \cdot f(0.25, 1.734) = 0.3968,$$

$$m_4 = (0.1) \cdot f(0.3, 1.9348) = 0.447.$$

Thus

$$y_3 = \tfrac{1}{6}(0.3476 + 0.785 + 0.794 + 0.447) = 1.934.$$

We know from Exercise 1(a) that the exact value of $y(0.3)$ is 1.933. The approximation is off by about 0.05 percent.

(b) Of course $x_0 = 0$, $y_0 = 1$. We calculate that

$$m_1 = (0.1) \cdot f(0, 1) = (0.1) \cdot f(0, 1) = 0.1,$$

$$m_2 = (0.1) \cdot f(0.05, 1.05) = 0.0952,$$

$$m_3 = (0.1) \cdot f(0.05, 1.0476) = 0.0955,$$

$$m_4 = (0.1) \cdot f(0.1, 1.0955) = 0.09128.$$

Thus

$$y_1 = 1 + \tfrac{1}{6}(0.1 + 0.1904 + 0.191 + 0.09128) = 1.0954.$$

Now $x_1 = 0.1$, $y_1 = 1.0954$. We calculate that

$$m_1 = (0.1) \cdot f(0.1, 1.0954) = 0.09129,$$

$$m_2 = (0.1) \cdot f(0.15, 1.14105) = 0.0876,$$

$$m_3 = (0.1) \cdot f(0.15, 1.1392) = 0.08778,$$

$$m_4 = (0.1) \cdot f(0.2, 1.18318) = 0.08452.$$

Thus

$$y_2 = 1.232 + \tfrac{1}{6}(0.09129 + 0.1752 + 0.17556 + 0.08452) = 1.1832.$$

Now $x_2 = 0.2$, $y_2 = 1.1832$. We calculate that

$$m_1 = (0.1) \cdot f(0.2, 1.1832) = 0.08452,$$

$$m_2 = (0.1) \cdot f(0.25, 1.22596) = 0.08157,$$

$$m_3 = (0.1) \cdot f(0.25, 1.2245) = 0.08167,$$

$$m_4 = (0.1) \cdot f(0.3, 1.2654) = 0.07903.$$

Thus

$$y_3 = 1.1832 + \tfrac{1}{6}(0.08448 + 0.16314 + 0.16334 + 0.7903) = 1.265.$$

We know from Exercise 1(b) that the exact value of $y(0.3)$ is 1.265. The approximation is precisely accurate to three decimal places.

5. Of course $x_0 = 0$, $y_0 = 2$. We calculate that

$$y_1 = 2 + 0.01 f(0, 2) = 1.96,$$

$$y_2 = 1.96 + 0.01 f(0.1, 1.96) = 1.9209.$$

The initial value problem $y' = x - 2y$, $y(0) = 2$, may be solved explicitly with solution $y = x/2 - 1/4 + [9/4]e^{-2x}$. We see then that the exact value of $y(0.3)$, to three decimal places, is $y(0.3) \approx 1.922$. The approximation is off by about 0.05 percent.

6. Of course $x_0 = 0$, $y_0 = 2$. We calculate that

$$z_1 = 2 + 0.01 f(0, 2) = 1.96,$$

$$y_1 = 2 + 0.005[f(0, 2) + f(0.01, 1.96)] = 1.96045,$$

$$z_2 = 1.96045 + 0.01 f(0.01, 1.96045) = 1.92134,$$

$$y_2 = 1.96045 + 0.005[f(0.01, 1.96045) + f(0.02, 1.92134)] = 1.922.$$

Comparing to the exact answer (to three decimal places) from Exercise 5, we see that the improved Euler method gives the precise answer to three decimal places.

7. Of course $x_0 = 0$, $y_0 = 2$. We calculate that

$$m_1 = 0.01 f(0, 2) = -0.04,$$

$$m_2 = 0.01 f(0.005, 1.98) = -0.03955,$$

$$m_3 = 0.01 f(0.005, 1.98023) = -0.03955,$$

$$m_4 = 0.01 f(0.01, 1.96045) = -0.039009.$$

It follows that

$$y_1 = 2 + \tfrac{1}{6}[-0.04 - 0.0791 - 0.0791 - 0.039009]$$

$$= 2 - 0.03993 = 1.96046.$$

Next we calculate y_2. Now

$$m_1 = 0.01 f(0.01, 1.96046) = -0.0391,$$

$$m_2 = 0.01 f(0.01, 1.94091) = -0.0387,$$

$$m_3 = 0.01 f(0.01, 1.94111) = -0.03872,$$

$$m_4 = 0.01 f(0.02, 1.92174) = -0.03823.$$

It follows that

$$y_2 = 1.96046 + \tfrac{1}{6}[-0.0391 - 0.0774 - 0.07744 - 0.03823]$$

$$= 1.9218.$$

Comparing with the exact answer (to three decimal places) in Exercise 5, we see that this solution is also accurate to three decimal places.

Chapter 8

1. (a) $\begin{cases} y' = z \\ z' = xz + xy. \end{cases}$

   (b) $\begin{cases} y' = z \\ z' = w \\ w' = w - x^2 z^2. \end{cases}$

   (c) $\begin{cases} y' = z \\ z' = xz + x^2 y. \end{cases}$

2. (a) We check that

$$4e^{4t} = \frac{d}{dt}(e^{4t}) = e^{4t} + 3e^{4t}$$

and

$$4e^{4t} = \frac{d}{dt}(e^{4t}) = 3e^{4t} + e^{4t},$$

hence the first pair is a solution of the system.

Likewise, we check that

$$-2e^{-2t} = \frac{d}{dt}(e^{-2t}) = e^{-2t} + 3(e^{-2t})$$

and

$$2e^{-2t} = \frac{d}{dt}(-e^{-2t}) = 3e^{-2t} + (-e^{-2t}),$$

hence the second pair is a solution of the system.

   (b) Set

$$X(t) = Ae^{4t} + Be^{-2t},$$

$$Y(t) = Ae^{4t} - Be^{-2t}.$$

The initial conditions give

$$5 = A + B$$

$$1 = A - B.$$

It follows that $A = 3$, $B = 2$. Hence the particular solution we seek is

$$X = 3e^{4t} + 2e^{-2t}$$

$$Y = 3e^{4t} - 2e^{-2t}.$$

3. (a) For the first pair, we check that

$$8e^{4t} = \frac{dx}{dt} = 2e^{4t} + 2 \cdot 3e^{4t}$$

$$12e^{4t} = \frac{dy}{dt} = 3 \cdot 2e^{4t} + 2 \cdot 3e^{4t}.$$

For the second pair, we check that

$$-e^{-t} = e^{-t} + 2(-e^{-t})$$

$$e^{-t} = 3(e^{-t}) + 2(-e^{-t}).$$

(b) We calculate that

$$3 = \frac{dx}{dt} = (3t - 2) + 2(-2t + 3) + t - 1$$

$$-2 = \frac{dy}{dt}3(3t - 2) + 2(-2t + 3) - 5t - 2.$$

The general solution is then

$$X = (3t - 2) + 2Ae^{4t} + Be^{-t}$$

$$Y = (-2t + 3) + Ae^{-t} + B(-e^{-t}).$$

4. (a) We set

$$0 = \det \begin{pmatrix} -3 - m & 4 \\ -2 & 3 - m \end{pmatrix} = m^2 - 1,$$

hence $m = \pm 1$. When $m = 1$ we have the algebraic system

$$-4A + 4b = 0$$

$$-2A + 2B = 0$$

with solutions $A = 1$, $B = 1$. This leads to the solution set $x = e^t$, $y = e^t$ for the system of differential equations. When $m = -1$ we have the algebraic system

$$-2A + 4B = 0$$

$$-2A + 4B = 0$$

with solutions $A = 2$, $B = 1$. This leads to the solution set $x = 2e^{-t}$, $y = e^{-t}$.

(b) This system is uncoupled. The solution of the first equation is $x(t) Ae^{2t}$ and the solution of the second equation is $y(t) = Be^{3t}$.

5. (a) $\begin{cases} y' = z \\ z' = w \\ w' = -x^2 w + xz - y + x \end{cases}$

   (b) $\begin{cases} y' = z \\ z' = (\sin x)z - (\cos x)y \end{cases}$

6. (a) We check that

$$-\tfrac{5}{2} Ae^{-5t} + 2Be^t = \frac{dx}{dt} = 3\left(\tfrac{1}{2} Ae^{-5t} + 2Be^t\right) - 4\left(Ae^{-5t} + Be^t\right)$$

and

$$-5Ae^{-5t} + Be^t = \frac{dy}{dt} = 4\left(\tfrac{1}{2} Ae^{-5t} + 2Be^t\right) - 7\left(Ae^{-5t} + Be^t\right),$$

hence this solution set satisfies the system of differential equations.

(b) We check that

$$3Ae^{3t} - Be^{-t} = \frac{dx}{dt} = \left(Ae^{3t} + Be^{-t}\right) + \left(2Ae^{3t} - 2Be^{-t}\right)$$

and

$$6Ae^{3t} + 2Be^{-t} = \frac{dy}{dt} = 4\left(Ae^{3t} + Be^{-t}\right) + \left(2Ae^{3t} - 2Be^{-t}\right),$$

hence this solution set satisfies the system.

7. (a) We calculate that

$$0 = \det \begin{pmatrix} 3 - m & 2 \\ -2 & -1 - m \end{pmatrix} = m^2 - 2m + 1,$$

hence we find the single root $m = 1$. We then solve the system

$$2A + 2B = 0$$

$$-2A - 2B = 0$$

to find that $A = 1$, $B = -1$, hence there is the solution set $x = e^t$, $y = -e^t$.

For a second solution, we guess

$$x = (\alpha + \beta t)e^t$$

$$y = (\gamma + \delta t)e^t.$$

Substituting into the system and solving yields $\alpha = 1$, $\beta = -1$, $\gamma = -1$, $\delta = 1$. Thus we find the second solution $x = (1 - t)e^t$, $y = (-1 + t)e^t$.

(b) We calculate that

$$0 = \det \begin{pmatrix} 1 - m & 1 \\ -1 & 1 - m \end{pmatrix} = m^2 - 2m + 2,$$

hence we find the roots $m = 1 \pm i$. Solving for A and B as usual, we find the solution sets

$$x = -ie^{(1+i)t}, \qquad y = e^{(1+i)t}$$

and

$$x = e^{(1-i)t}, \qquad y = -ie^{(1-i)t}.$$

Taking a suitable linear combination of these two complex-valued solutions, we find the real solutions

$$x = e^t \cos t, \qquad y = -e^t \sin t$$

and

$$x = e^t \sin t, \qquad y = e^t \cos t.$$

8. We calculate that

$$0 = \det \begin{pmatrix} 0 - m & 1 \\ 1 & 0 - m \end{pmatrix} = m^2 - 1,$$

hence we find the roots $m = \pm 1$. Solving for A and B as usual gives the solution sets

$$x = e^t, \qquad y = e^t$$

and

$$x = e^{-t}, \qquad y = -e^{-t}.$$

The general solution is

$$x = Ae^t + Be^{-t}, \qquad y = Ae^t - Be^{-t}.$$

The initial conditions yield the equations

$$1 = A + B$$

$$0 = Ae - Be^{-1}.$$

Solving gives us the particular solution

$$x = \frac{1}{1 + e^2}e^t + \frac{e^2}{1 + e^2}e^{-t}, \qquad \frac{1}{1 + e^2}e^t - \frac{e^2}{1 + e^2}e^{-t}.$$

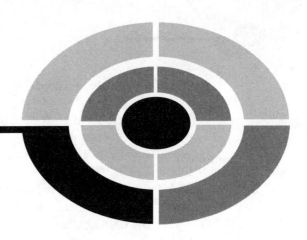

Bibliography

[ALM]   F. J. Almgren, *Plateau's Problem: An Invitation to Varifold Geometry*, Benjamin, New York, 1966.

[BIR]   G. Birkhoff and G.-C. Rota, *Ordinary Differential Equations*, Ginn & Co., New York, 1962.

[BLI]   G. A. Bliss, *Lectures on the Calculus of Variations*, University of Chicago Press, Chicago, 1946.

[BRM]   R. Brooks and J. P. Matelski, The dynamics of 2-generator subgroups of PSL(2, $\mathbb{C}$), *Riemann Surfaces and Related Topics*, Proceedings of the 1978 Stony Brook Conference, pp. 65–71, *Ann. of Math. Studies* 97, Princeton University Press, Princeton, NJ, 1981.

[COL]   E. A. Coddington and N. Levinson, *Theory of Ordinary Differential Equations*, McGraw-Hill, New York, 1955.

[FOU]   J. Fourier, *The Analytical Theory of Heat*, G. E. Stechert & Co., New York, 1878.

[GAK]   T. Gamelin and D. Khavinson, The isoperimetric inequality and rational approximation, *Am. Math. Monthly* 96(1989), 18–30.

[GER]   C. F. Gerald, *Applied Numerical Analysis*, Addison-Wesley, Reading, MA, 1970.

[HIL]   F. B. Hildebrand, *Introduction to Numerical Analysis*, Dover, New York, 1987.

[ISK]   E. Isaacson and H. Keller, *Analysis of Numerical Methods*, John Wiley, New York, 1966.

[JOL]   J. Jost and X. Li-Jost, *Calculus of Variations*, Cambridge University Press, Cambridge, 1998.

[KRA1]   S. G. Krantz, *Complex Analysis: The Geometric Viewpoint*, 2nd ed., Mathematical Association of America, Washington, DC, 2003.

[KRA2]   S. G. Krantz, *Real Analysis and Foundations*, CRC Press, Boca Raton, FL, 1992.

[KRA3]   S. G. Krantz, *A Panorama of Harmonic Analysis*, Mathematical Association of America, Washington, DC, 1999.

[KRA4]   S. G. Krantz, *The Elements of Advanced Mathematics*, 2nd ed., CRC Press, Boca Raton, FL, 2002.

[KRP1]   S. G. Krantz and H. R. Parks, *A Primer of Real Analytic Functions*, 2nd ed., Birkhäuser, Boston, 2002.

[KRP2]   S. G. Krantz and H. R. Parks, *The Implicit Function Theorem*, Birkhäuser, Boston, 2002.

[LAN]   R. E. Langer, *Fourier Series: The Genesis and Evolution of a Theory*, Herbert Ellsworth Slaught Memorial Paper I, *Am. Math. Monthly* 54(1947).

[LUZ]   N. Luzin, The evolution of "Function," Part I, Abe Shenitzer, ed., *Am. Math. Monthly* 105(1998), 59–67.

[MOR]   F. Morgan, *Geometric Measure Theory: A Beginner's Guide*, Academic Press, Boston, 1988.

[OSS]   R. Osserman, The isoperimetric inequality, *Bull. AMS* 84(1978), 1182–1238.

[RUD]   W. Rudin, *Functional Analysis*, 2nd ed., McGraw-Hill, New York, 1991.

[STA]   P. Stark, *Introduction to Numerical Methods*, Macmillan, New York, 1970.

[STE]   J. Stewart, *Calculus: Concepts and Contexts*, Brooks/Cole Publishing, Pacific Grove, CA, 2001.

[THO]   G. B. Thomas (with Ross L. Finney), *Calculus and Analytic Geometry*, 7th ed., Addison-Wesley, Reading, MA, 1988.

[TIT]   E. C. Titchmarsh, *Introduction to the Theory of Fourier Integrals*, The Clarendon Press, Oxford, 1948.

[TOD]   J. Todd, *Basic Numerical Mathematics*, Academic Press, New York, 1978.

[WAT]   G. N. Watson, *A Treatise on the Theory of Bessel Functions*, 2nd ed., Cambridge University Press, Cambridge, 1958.

[WHY]   G. Whyburn, *Analytic Topology*, American Mathematical Society, New York, 1948.

INDEX

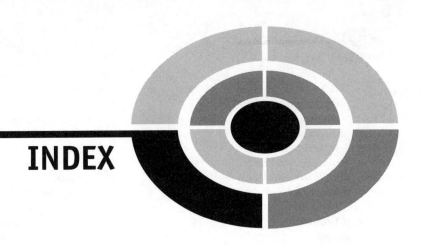

Abel, N. H., 184
   mechanical problem, 184
   problem of a bead sliding down a wire, 184
addition of series, 100
alternating series test, 95
analogy between electrical current and flow of
    water, 44
approximate solution graph, 199
approximation by parabolas, 214
associated
   linear algebraic system, 225
   polynomial, 49

Bernoulli, D., 143
   solution of wave equation, 143
Bessel functions, 93, 179
binary recursion, 113
boundary
   conditions, 144
   value problem, 144, 162
bounded variation
   function as the difference of two monotone
     functions, 128
   variation, functions of, 128
brachistochrone, 39, 188
Brahe, Tycho, 85

capacitor, 43
catenary, 39
Cauchy, A. L., 144
   product, 100
   product of conditionally convergent series, 101

product of series, 100
–Schwarz–Bunjakovski inequality, 137
Cesàro means, 127
complex
   exponentials, 51
   numbers, 53
   roots for higher-order equations, 87
condenser, 43
conservation of energy, 151, 185
constants in solutions, 4
convergence of Fourier series, 125
convolution, 180
   and the Laplace transform, 180
cosine series expansion, 131
coupled harmonic oscillators, 89

d'Alembert, J. L., 142
   solution of vibrating string, 142
damped vibrations, 68
damping is less than force of spring, 70
density of heat, 151
derivative of the Laplace transform, 176
descent time of a sliding bead, 185
differential equation
   examples of, 3
   in physics, 2
   that describes a family of curves, 19
   use of to derive power series, 103
   what is, 1
Dirichlet, P. L., 125, 156
   and convergence of series, 144

Dirichlet, P. L. (*contd.*)
  conditions, 128
  problem for a disc, 156, 159
discontinuity
  of the first kind, 125
  of the second kind, 125
discrete models, 199
discretization error, 204
  local, 205
  total, 206
distinct
  complex roots for systems, 228
  real roots for higher-order equations, 87
  real roots for systems, 226
double precision calculations, 205

eigenfunctions, 145, 162
eigenvalues, 145
electrical
  circuits, 43, 74
  flow analogous to oscillating cart, 74
electromagnetics, 156
electromotive force, 43
elliptic equation, 148
end of solution process, 5
error
  estimates, 205
  terms, 203
estimate for discretization error, 205
Euler, L., 116
  equidimensional equation, 158
  formula, 52
  method, 201
  method, improved, 207
  method, rationale for, 201
even
  extension of a function, 131
  functions, 128
exact equations, 13
  method of, 14
exactness and geometry, 17
existence and uniqueness for systems, 221

falling body, 2
  described by a system, 233
Fejér, L., 126

filters, 121
first-order
  equations, solution with power series, 102
  linear equations, 10
  method of, 10
forced vibrations, 72
Fourier, J., 144
  book, 154
  coefficients on an arbitrary interval, 133
  derivation of the formula for Fourier
    coefficients, 154
  series, 115
  series, applications of, 115
  series, coefficients of, 116
  series, convergence of, 124
  series, mathematical theory, 144
  series on arbitrary intervals, 132
  series on $[-L, L]$, 133
  series summands, 119
  series, uniform convergence of, 127
  series vs. power series, 115
  solution of the heat equation, 151
foxes and rabbits, 233
friction
  exceeds string force, 69
  piecewise smooth, 125
  what is, 143

Gauss, C., 144
general solution, 5, 31, 32, 50
  of a system, 223
geometry of 3-space, 136
Golden Gate Bridge, 74

Halley, Edmund, 85
hanging chain problem, 36
heat distribution on a disc, 158
heat equation, 152
  derivation of, 151, 152
  Fourier's point of view, 151
  Fourier's solution of, 153
heat flow
  in a disc, 156
  in a rod, 156
heat has no sources or sinks, 152
heated rod, 151

Heun's method, 208
higher-order
    differential equations, 85
    linear equations, solution of, 86
higher transcendental functions, 93
homogeneous, 22
    equation, 22, 49, 55
    of degree α, 23
    systems, 219, 221
    systems with constant coefficients,
        225
hyperbolic partial differential equation,
    148

imaginary numbers, 52
improved Euler method, 207, 208
impulse, 193
    function, 192
impulsive response, 194
independent solutions, 7
indicial response, 189
inductance, 43
infinite-dimensional spaces, 136
initial conditions, 34
inner product, 137
input equal to a step function, 191
integral
    equations and the Laplace transform,
        183
    of the Laplace transform, 179
integrating factors, 17, 26
interval of convergence, 94
    endpoints of, 95
inverse Laplace transform, 173

Kepler, J., 75
    and Tycho Brache, 85
    First Law, 75
    First Law, derivation of, 80
    laws, 75
    Second Law, 75
    Second Law, derivation of, 78
    Third Law, 75
    Third Law, derivation of, 82
Kirchhoff's Law, 44
known solution, use of to find another, 62, 63

Lagrange, J. L., 144
    interpolation, 144
Laplace, P., 156
Laplace transform, 168
    analysis of Bessel's equation, 177
    calculation of, 170
    converting a differential equation, 172
    definition of, 169
    is one-to-one, 173
    of the antiderivative, 175
    of the derivative, 171
    properties of, 180
    solving a differential equation, 172
laws of nature, 3
Legendre, A. M., 3
    equation, 109
    functions, 112
    polynomials, 112
length, 137, 138
linear combinations of solutions, 5, 222
linearization, 237
linearly
    dependent, 223
    independent, 224

Maple, 204
Mathematica, 204
method of linearization, 237
method of reduction of order, 30
method of wishful thinking, 27
monotone increasing, 128

n-body problem, 217
Newton's Law
    of Cooling, 151
    of Universal Gravitation, 75
Newtonian model of the universe, 219
noise and hiss, 121
nonlinear systems, 233
nonlinearity, 238
norm, 137, 138
numerical
    analysis, 198
    approximation of solutions, 198

numerical (*contd.*)
  method, spirit of, 199
  methods, 198

odd
  extension of a function, 131
  functions, 128
order of an equation, 3
organized guessing, 49, 60
orthogonal
  expansions, 151
  functions, 136
  system, 162
  trajectories, 20
orthogonality, 136
  of eigenfunctions, 150
  properties of eigenfunctions, 163
  property, 162
  with respect to a weight, 165
orthonormal, 162

parabolic equation, 148
parity relations, 129
particular solution, 34
periodicity, 119
physical principles governing heat, 151
Poisson, S. D., 159
  integral, 159, 161
  integral formula, 161
  kernel, 161
polynomials, 92
population ecology, 217
potential theory, 156
power series, 93
  convergence of, 93, 99
  convergence to a function, 98
  formula for coefficients, 97
  solution at an ordinary point, 107, 113
  sum of, 94
  uniqueness of, 98
  vs. Fourier series, 115
products of series, 100
pseudosphere, 41
pursuit curves, 40

radius of convergence, 94
ratio test, 94

real analytic functions, 98
  properties of, 102
reduction of order
  method of, 30
  with dependent variable missing, 30
  with independent variable missing, 32
remainder term in Taylor's formula, 98
repeated real roots
  for higher-order equations, 87
  for systems, 230
resistance balances force of spring, 69
resistor, 43
resonance, 74
response
  of an electrical system, 189
  of a mechanical system, 189
root test, 96
round-off error, 204
  dangers of, 204
Runge–Kutta method, 210, 212

scalar multiplication of series, 100
second-order equations, power series solution
    of, 107
second-order linear equations, 48
  with constant coefficients, 48
separable equations, 7
  method of, 8
separable, not all equations are, 10
separation of variables, 148, 157
series
  operations on, 100
  products of, 100
  sums of, 100
simple discontinuity, 125
simple harmonic motion, 66
  undamped, 66
Simpson's rule, 210
sine series expansion, 131
smaller step size, 210
solution
  as an implicitly defined function, 5
  expressed implicitly, 18
  has no derivatives, 6
  of a differential equation, 4
  qualitative properties, 198
  set for a system, 221

special functions, 93
square-integrable functions, 139
steady-state heat distribution, 155
step function, 189
step size too small, 204
Sturm–Liouville
    problems, 162
    theory, 163
sums
    of series, 100
    of solutions of Laplace's equation, 158
superposition, 5
    formulas, 195
system as a single vector-valued equation, 223
systems
    as vector-valued differential equations, 217
    of differential equations, 216
    of linear equations, 219

tautochrone, 39, 188
Taylor expansions, 98
Taylor's formula, remainder term in, 99
tractrix, 40, 41
transcendental functions, 92
transforms, 168
Treatise on the Theory of Heat, 154
Triangle inequality, 137
trigonometric series, 115, 116
two-term recursion, 113

uncoupled systems, 230
underdamped system with forcing
    term, 73
undetermined
    coefficients, 54
    coefficients for higher-order
        equations, 58
    coefficients with repeated roots, 56
    constants in solutions, 7
universal law of gravitation, 85
Uranus, orbit of, 84

variation of parameters, 58
    method of, 59
vector space, 136
vibrating string, 141, 145
vibrations and oscillations, 65
Volterra, V., 233
    method of, 235
    predator–prey equations, 234

wave equation, 142
    derivation of, 145, 146
    solution of, 148
well-posed problem, 142
Wronskian determinant, 164
Wronskian for a system, 223

ABOUT THE AUTHOR

Steven Krantz, Ph.D., is Chairman of the Mathematics Department at Washington University in St. Louis. An award-winning teacher and author, Dr. Krantz has written more than 45 books on mathematics, including *Calculus Demystified*, another popular title in this series. He lives in St. Louis, Missouri.